AF449380

*Rubber Seals for Fluid
and Hydraulic Systems*

Dedication

This book is dedicated

to Arputhamary my wife,

Peter Premkumar

and

Arunkumar my sons.

Rubber Seals for Fluid and Hydraulic Systems

V.C. Chandrasekaran

AMSTERDAM • BOSTON • HEIDELBERG • LONDON • NEW YORK
OXFORD • PARIS • SAN DIEGO • SAN FRANCISCO
SINGAPORE • SYDNEY • TOKYO

William Andrew is an imprint of Elsevier

William Andrew is an imprint of Elsevier
The Boulevard, Langford Lane, Kidlington, Oxford OX5 1GB, UK
30 Corporate Drive, Suite 400, Burlington, MA 01803, USA

First edition 2010

Notice
No responsibility is assumed by the publisher for any injury and/or damage to persons or property as a matter of products liability, negligence or otherwise, or from any use or operation of any methods, products, instructions or ideas contained in the material herein. Because of rapid advances in the medical sciences, in particular, independent verification of diagnoses and drug dosages should be made

British Library Cataloguing in Publication Data
A catalogue record for this book is available from the British Library

Library of Congress Cataloging-in-Publication Data
A catalog record for this book is availabe from the Library of Congress

ISBN–13: 978-0-8155-2075-7

For information on all Elsevier publications
visit our web site at books.elsevier.com

Printed and bound in the United States of America

10 10 9 8 7 6 5 4 3 2 1

Working together to grow
libraries in developing countries

www.elsevier.com | www.bookaid.org | www.sabre.org

ELSEVIER   BOOK AID International   Sabre Foundation

Contents

Preface

In writing this book, I aim to summarize the subject matter, placing an emphasis on important areas such as applications of rubber as fluid seals in the nuclear, aviation, oil drilling and automotive industries.

The principal readership of the book will be chemists in the discipline of rubber science and technology; engineers in the field of chemical, mechanical, nuclear, oil field and aviation engineering; teachers, students, research fellows and practicing professionals in several fields of science and technology and last, but not least, all the buying and marketing professionals in various engineering sectors.

A chapter on Rubber Expansion Joints is included since the author feels that the function of such expansion joints as pipe connectors is indirectly linked with leakage and prevention of fluid flow through pipes.

In Chapter 10, 'Manufacture of Seals and 'O' Rings', which is the longest chapter, about 25 workable starting point formulations based on different rubbers are included, with cure and property data for those formulations as guidelines for technologists and engineers.

It is assumed that the readers have a sound working knowledge of the relevant chemistry, physics and engineering, as well as being well versed in the general terminology of the subject.

Almost throughout this book, the term 'rubber' is used, and wherever 'polymer' or 'elastomer' are substituted, they carry the same meaning as rubber unless otherwise explained, since these terms are synonyms.

Repetitions here and there are not mistakes, but are included with the intention of emphasizing a particular point or providing necessary contextualization.

I would like to invite comments and suggestions from all engineers, chemists and technologists who are concerned with fluid leakage problems, so that I can continue to develop and improve this work in future editions.

Acknowledgements

While writing this book, the author could not forget the assistance received from Priya Victor, Malathy Peter and Eswari Arun, all of whom were enthusiastic in motivating me and updating my office infrastructure as well. The invaluable suggestions and criticisms provoked me, and enabled me to approach each topic with vigilance and clarity as far as possible.

Vast resources of information in the form of abstract and full papers by leading authors were provided by the Rubber and Plastics Research Association, UK, were greatly helpful to me. I thank them with immense gratitude.

Lastly, Vankatakrishnan Ranganathan, who spent all day with me in front of the computer during the preparation of the manuscript, is in my mind as usual and I thank him too.

Notice (disclaimer)

The information on fluid seals and the related subject matter dealt with in this book is given in good faith and not with any warranty. The information is based on the exposure and general experience of the author. Because several factors that are out of the author's knowledge and control can affect the use of this information, any injury, loss or damage resulting from relying on such information cannot be considered as our or the publisher's liability.

CHAPTER 1

INTRODUCTION

Seals prevent escape of fluids from a hollow cylinder when a shaft penetrates the wall of the cylinder. Most commonly, the shaft will have a rotary or linear movement. If a seal is not made as per functional requirements, or installed and maintained properly, it may fail, causing loss of fluid. The two main functions of a seal are to keep the fluid in while keeping dirt and debris out.

Fluid leaks in hydraulic and lubricating systems are a commonplace occurrence in much mechanical equipment, and they are undesirable and embarrassing events fuelling increased cost of any manufacturing and equipment maintenance activity. Often fire hazards and accidents are direct consequences of such leakages. In many situations, in spite of anticipation of such hazards, leaks are overlooked and the leaking system is not attended to, since the operating engineers do not take the threat of accidents seriously. For years, leakage has received low level attention by corporate management at many industrial manufacturing plants. One reason is that leaks are often viewed as inevitable, and corrective actions such as repair and replacement are done only when convenient.

According to a recent report by John C Cox, Business Development Manager of Swagelok Company Solon Ohio [1], fluid leakage costs industry millions of dollars every year. For example, a few small leaks in a facility using air at 100 psig with an electric consumption cost of 6 cents/kwh can waste more than $22 000 annually. Delaying the replacement of a leaking steam trap worth $100, or not repairing it with proper sealing, could waste $50 per week. Since an average facility can typically have hundreds of steam traps throughout its operations, leaking traps may be squandering hundreds of thousands of dollars each year. In addition to wasted dollars, unattended leaks can result in downtime, affect product quality, pollute the environment and cause injury. System vibration, pulsation and thermal cycling are all common causes of chemical processing system leakage. It can be assumed that any type of fitting connection may leak, regardless of whether pipe or tube is used, especially when mechanical vibration is present. This 'vibration fatigue' is an unavoidable factor that can be aggravated by poor metallurgical consistency in the construction material, undue stress imposed on the connection from side load, other system design characteristics, or simply improper installation practices.

Rubber Seals for Fluid and Hydraulic Systems 2010; ISBN: 9780815520757

SEAL FAILURES AND FLUID LEAK

The increasing speed of mechanical systems, warranted by the desire for increased productivity, leads to higher operating temperatures and reduced fluid viscosities. This, associated with higher pressures, causes an increasing tendency for fluids to leak. Such leakage in fuel systems that handle highly inflammable solvents cannot be overlooked as there is a high probability of fire hazards. It has become common practice to include a safe leakage route in the system design, to an escape or collection point in order to minimize the risk. However, this is a wasteful solution both in terms of cost and additional space needed for such a device. Unattended leaks result in downtime, affect product quality and pollute the atmosphere, in addition to wasting dollars.

If a seal leaks for no apparent reason, or shows cracks and blisters, it could be reacting with the fluid it is sealing. This problem is especially severe if the fluid has a molecular structure similar to that of the seal rubber. If this fluid interaction is the cause of seal failure, one has to see how to avoid the problem.

Blisters and internal cracks indicate that a seal is interacting with a fluid. Such failures, which are evident only upon close examination of the seal, are caused by fluid permeating into and expanding microscopic voids in the rubber.

Rubber seal materials are normally chosen on the basis of chemical compatibility with the fluid being transferred, but even if seal and fluid are chemically compatible, they can still interact physically, leading to leakage. Failure results from the fluid literally digging into the seal and bursting it from the inside. The reaction produces blisters or cracks in the seal that eventually provide a path for fluid leakage. This problem is especially acute at high temperatures and pressures, but it can show up even under less severe operating conditions. Work with seals for oil wells has provided insight on how to recognize failures caused by fluid/seal interaction. Fluids interact with seals because no seal material is 100% solid. Typically, molecular voids or air spaces make up about 3% of a seal's volume at its glass-transition temperature [2]. Under pressure, fluid diffuses into these voids and reaches an equilibrium state. Any change in equilibrium creates a positive internal pressure in the voids. If the pressure exceeds a critical level, the voids expand, forming a blister or rupture. Blistering is characteristic of highly elastic materials used to seal supersaturated gases, that have the following physical properties:

- low hardness (low shear modulus)

- low cross-link density

- high elongation ($>200\%$) [3].

An example of leak prevention with a rubber seal in a thread fitting that is gaining popularity is the Society of Automotive Engineers' (SAE) straight thread. The SAE straight threads are

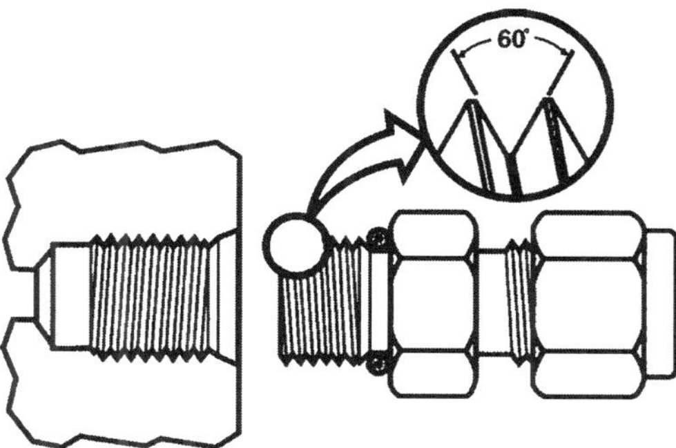

Figure 1.1: SAE straight-thread fittings – straight threads are mechanical types; they do not provide a seal

mechanical types, designed to only hold the fitting in place; they do not provide a seal. This is provided by a rubber inset, typically located at the base of the male thread (Figure 1.1). The rubber compresses against a flat surface near the entrance to the female port. This type of threaded seal offers the advantage that the maintenance, accessibility and replacement of the fitting is significantly easier for the installer.

In another interesting report, the principal author, S. Hunt of Dominion Engineering Inc., states the following in one of his chapters [4]:

> Leakage has a number of significant consequences including spread of contamination, pronounced hazard, damage to equipment, loss of efficiency and power reduction. For average plants the annual cost of leakage is in the range of \$850 000 to \$2.4 millions per unit. A proactive approach to reducing rather than responding to the leaks has the potential to reduce leakage related costs in savings of \$200 000 to \$500 000 per unit per year.

Successful sealing involves containment of fluid within the system while excluding contaminants. For example, in a typical reciprocating sealing system, rubber materials are found to have dimensional variations due to manufacturing tolerances, loads and deformations in the cylinder under pressure. System pressure on the seal surface attempts to compress the seal axially. This compression forces the seal more tightly against its contacting metal surface. The resilience of rubber creates a tight sealing effect containing the fluid and excluding any contaminants.

The mechanical engineer who is familiar with metals looks to a rubber technologist whenever leakage in a fluid system is encountered. Leakages are encountered in a range of disciplines, such as fluid flow, lubrication, hydraulic system and piping technologies. The leak needs to be rectified for at least the life-time of the component where the seal is fitted. Mechanical engineers involved in many branches of industry are more familiar with metals rather than elastic rubbers. Often they turn to rubber technologists, since rubber is a unique and appropriate material for sealing fluid leaks unlike rigid metals. 'What materials make a good seal?' asks Professor Michael F. Ashby and he answers 'Elastomers – everyone knows that' [5]. The terms elastomer and rubber are scientifically interchangeable, although the latter is used in some areas to refer only to natural rubber, which comes from the latex of hevea trees, as opposed to synthetic rubber, which is generally a petroleum oil by-product. Some standards use the term elastomer for cross-linked materials, but there is no general agreement on this. In the literature, the terms elastomer, polymers and rubbers are synonymous.

How can a rubber seal be explained simply? It can be visualized as a cylindrical item compressed between two flat surfaces as shown in Figure 1.2.

The seal must form the largest possible contact width 'b' and must not damage the contact surfaces. The seal itself must remain elastic so that it can be dynamic or static.

Although the author enumerates some application details of rubber seals, the primary focus of this book is on the characteristics of rubbers as seals. Their manufacturing procedures, the implications of their physical and chemical characteristics for sealing in fluid and hydraulic systems, how rubbers seal and prevent leaks, what properties are required for sealing function, and how they change before and after installation are also discussed. It is hoped that this will draw mechanical engineers closer to rubber technology.

The author has experience of working with mechanical, chemical, aviation and ship building engineers to solve their leakage problems, it is hoped, therefore, that this book will benefit those people, as well as scientists and technologists engaged in leakage prevention management. The author believes that this book might contribute its share in combating fluid leak threats. During his search, he found no book that dealt exclusively with rubber as industrial fluid seals was available in the literature, although valuable information related to the subject

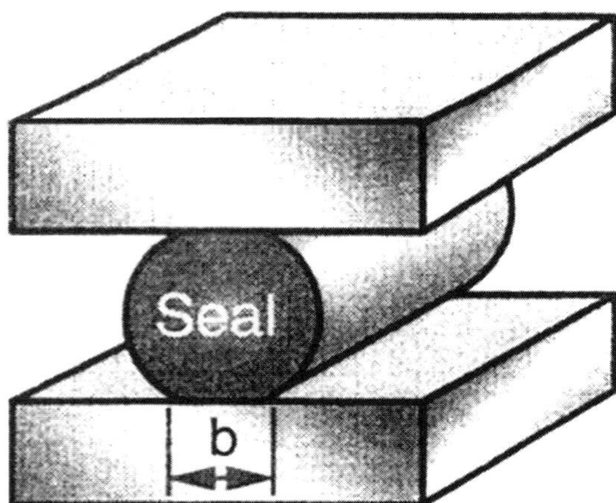

Figure 1.2:   Rubber compressed between two flat surfaces

could be found in industry documents and research reports. These are cited where relevant, and detailed in the Reference Sections. A sound understanding of the complex factors involved in successful fluid sealing is essential for engineers who specify, design, operate and maintain machinery and mechanical equipment. Rubber seal specialists will show how an understanding of basic engineering factors can be used to practical advantage. Fluid sealing technology is based on principles as diverse as lubrication, friction, wear, properties of materials, mechanical design, fluid mechanics and heat transfer. All of these factors are to be considered and studied in depth in designing different types of rubber seals, seal materials and sealing applications.

REFERENCES

1 www.swagelok.com/downloads/webcatalogs
2 Hertz, D.L. Jr Sour hydrocarbons – the elastomer challenge. Meeting of the Rubber Div., American Chemical Society, Las Vegas, Nev., May 20–23, 1980.
3 Hertz, D.L. Jr The hidden cause of seal failure. Machine design, April 9, 1981.
4 http://mydocs.epri.com/docs/public Establishing an effective fluid leak management program: Sealing technology and plant leakage reduction series. Electric Power Research institute (EPRI), Palo Alto, CA 2000, Technical report no: 114761, 2002.
5 Ashby, M.F. Material selection in mechanical design, 3rd edn. Elsevier Butterworth, Oxford, 2005.

CHAPTER 2

RUBBER PROPERTIES FOR FUNCTIONAL SEAL REQUIREMENTS

Rubber seals are used in a range of applications, in which the forces that operate on them are very different. To optimize the performance of a seal, the forces that operate in a given application, together with the functional properties that are needed to respond to them, need to be understood.

When a rubber seal is used in an external application, such as an 'O' ring, it is usually stretched over the piston into a groove where, in general, it will initially be at a maximum stretch of 8%. The sectional diameter of the ring will become reduced because of this stretch and, when the piston is assembled into the cylinder bore, this section will be compressed giving a sealing effect due to the recovering force of the rubber. Stress relaxation will occur when the seal is under constant strain for a prolonged period and compression set will be observed. If this occurs in a reciprocating application on a piston head, the 'O' ring will be pushed back and forth in its housing and may extrude out on both sides and be twisted spirally.

Alternatively, squeeze type seals operate by distorting under compressive load, and the hardness specified for such an application must be sufficient to ensure adequate retention of the sealing pressure. The squeeze resistance can be enhanced by incorporating one or more plies of fabrics in the rubber section rather than increasing the hardness of the compound as this might adversely affect other properties. Dimensional variations due to contact with fluids can be adjusted to achieve a small positive swell which can maintain seal efficiency, by compensating for wear and compression set. The choice of compound will depend on the effects of the fluids with which the seal is in contact, the operating temperature, and mechanical conditions such as pressure, relative velocity and abrasion.

In the design of seals for specific applications, the engineer has first to understand the limitations of the physical properties of the rubber material to be used in order to avoid applying stresses (applied loads) and strains (deformations) that exceed these limits. The strengths of rubbers are considerably lower than those of metals, plastics or wood, but its elasticity is much greater. The rubber's ability to return to its original shape and dimensions after a deforming force is removed is often called resilience or memory. Resilience implies quick return, whilst

Rubber Seals for Fluid and Hydraulic Systems 2010; ISBN: 9780815520757

memory implies slow return. In seals, resilience is important because it allows a dynamic seal to adapt to variations in the sealing surface. Memory, which implies a slow return, is the same as creep.

The reversibility of elastic deformation of rubber is known as *hysteresis*. If, for a given elongation before breaking, the stress is relaxed and the sample is allowed to retract, the energy of retraction is lower than that used for stretching. This phenomenon is hysteresis [1]. Thus, hysteresis indicates that the energy recovered by the rubber during retraction is distinctly lower than the energy used in stretching. If a rubber sample is subjected to several successive stretches followed by retractions, it will be found that the rubber sample finally reaches a limit when the retracted length is longer than the original unstretched length. This is called *permanent set*. For vulcanized rubber, the permanent set is smaller, but is influenced by the compound formulation. Permanent set is a measure of elasticity of a rubber. In practice, permanent set also occurs during compressive stresses. For a seal application, the rubber has to decay or relax less, and the permanent set has to be as low as possible.

In the case of metals, the creep phenomenon manifests in a different way, as shown in Figure 2.1. For instance, when a vertical metallic bar is subjected to a load 'p' for a given period of time 't_0', it produces an elongation equal to 'δ_0'. Subsequent to time 't_0', if the load is kept constant and the time increased, the bar gradually and slowly lengthens. This lengthening effect, i.e. the increase of strain on the metallic bar whilst the load (stress) does not change, is called the creep behavior of the metal [2].

In a second example of creep, a wire is stretched between two immovable supports so that it has an initial tensile stress 'σ_0' (Figure 2.2). The time for the initial stretch is 't_0'. However, after a period of time, the stress in the wire gradually diminishes, eventually reaching a constant value, while the supports at the ends of the wire do not move. This process is called stress relaxation, or stress decay of the material.

The resistance to wear and abrasion of a rubber seal when in contact with a moving surface is related to other physical properties, such as hardness and tear resistance. Thermal

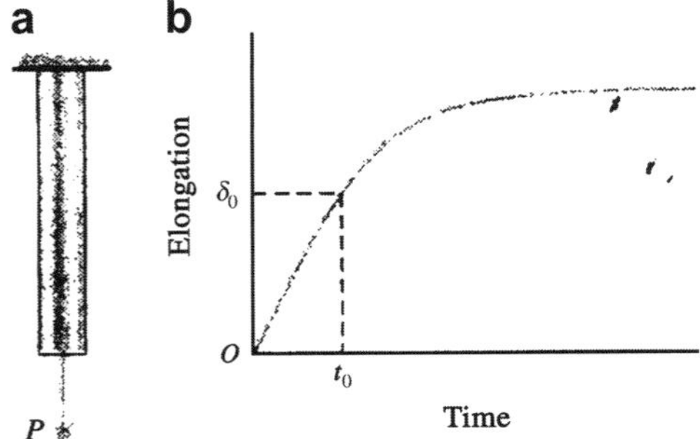

Figure 2.1: Creep in a bar under constant load

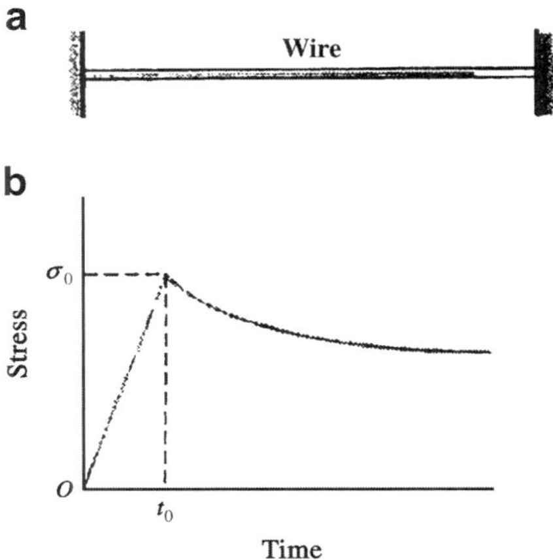

Figure 2.2: Relaxation of stress in a wire under constant strain

stability, fluid compatibility and fatigue resistance also are associated with wear resistance. Vibration and pulsation in mechanical systems, or pipelines are common in chemical process industries and can lead to leakages of corrosive fluid. The vibration fatigue is an unavoidable factor which is further aggravated by improper installation of the moving system. Leakages due to these pulsations can be prevented by the installation of flexible rubber expansion joints.

Chemical compatibility is another important factor to reckon with in seal design. For instance, butterfly valves are frequently used in industrial fluid flow systems to load and unload different kinds of fluids for shipment or storage. These applications include tank trailers for fluid transportation and fluid flow pipe lines. A typical butterfly valve usually incorporates a resilient seal against which the valve pivots to seal off the flow lines and, when opened, allows the fluid to be conveyed to the flow line. The chemicals transported in tanks using such valves are often caustic or acidic, and can corrode the pivot area of the valve. The valve may fail, even though the valve chamber is protected by a chemically resistant rubber material.

In iron ore mines, the concentrated ore is made as a slurry in the concentrator plant and transported to pelletizing plants via long distance pipe lines. These pipes, having no flanges, are joined with victaulic couplings, which are recent improvements, fixed on grooves in the pipe ends. These couplings accommodate rubber sealing gaskets to prevent leakage of slurry, whilst allowing material of the required density to enable flow through the pipes. The slurry is transported under high pressure to pelletizing plants and other destinations. The rubber ring gaskets installed in the coupling play a vital role in prevention of leakage of the slurry. In such cases, the rubber must withstand a tight sealing pressure and the surface in contact with the slurry, (which contains highly abrasive ore particles) should have high cut and tear resistance.

The resilient rubber provides a tight and durable seal. Generally, high pressure application facilitates sealing improvements. Rubber seals used under high pressure must have a high tear strength, hardness and modulus to prevent extrusion of the rubber gasket. It is usual practice for in this kind of pressure system, the rubber seal to be backed by a rubber of high hardness in order to prevent extrusion.

Seal manufactures develop their own rubber compounds suitable for seals, which possess the chemical, physical and swelling properties to match the functional requirements and working conditions of the application. The compounds used in the manufacture of seals are derived from base rubbers such as natural rubber, nitriles, neoprenes, butyls, styrene butadiene, carboxylated nitriles, viton, silicones and polytetrafluoroethylene. Of all the properties exhibited by the various types of rubber compounds, the most critical ones pertain to how they change when they are installed as seals and while in service. All physical properties change with age, and exposure to variations in temperature, fluid type, pressure, and other factors which can include corrosive chemicals and fumes and gases. Compounds with the smallest tendency to change their properties, whether chemical or physical, are easier to work with. More adaptable and versatile seals can be produced with these compounds.

LOW TEMPERATURE BEHAVIOR – GLASS TRANSITION TEMPERATURE (Tg)

A fundamental property of an elastomer is the *glass transition temperature* (Tg), which differs from one to another. For example, for natural rubber Tg is $-70°C$ ($-95°F$). Broadly, this means that, above $-70°C$, the material behaves as a rubber, but below $-70°C$, the material behaves more like glass. When glassy, natural rubber is about one thousand times as stiff as it is when rubbery. When glassy, a hammer blow on natural rubber will cause it to shatter like glass, whilst, when rubbery, the hammer will probably just bounce off.

At normal temperatures, the rubber chain molecules are in a constant state of thermal motion; they are constantly changing their configuration and this movement makes them reasonably easy to stretch. It is to be noted that, as the temperature is lowered, the chains become less flexible and the amount of thermal motion decreases. Eventually, at the glass transition temperature, all major motion of the chains ceases. The material no longer has the properties which make it rubber and it behaves as glass.

For all practical engineering uses of elastomers we require good flexibility and, therefore, it is essential that we use them only at temperatures which are comfortably above the glass transition. This is generally no problem for natural rubber with a Tg of $-70°C$, or cis-butadiene rubber with a Tg of $-108°C$ ($-160°F$). But many elastomers, especially those which have been designed to be highly heat or oil resistant, have much higher Tgs and this must be borne in mind when selecting them for service applications. For example, some fluoroelastomers, which have excellent oil and heat resistance, have a Tg not far below normal room temperature. This can result in problems if a component required to work at high temperature also needs to function under cooler conditions.

HIGH TEMPERATURE BEHAVIOR

The limit to the upper temperature at which a rubber can be used is generally determined by its chemical stability and will thus vary for different rubbers. Rubbers can be attacked by oxygen or other chemical agents and, because the attack results in a chemical reaction, degradation will increase with temperature.

Degradative chemical reactions are generally of two types. The first are those which cause breakage of the molecular chains or cross-links, softening the rubber because they weaken the network. The second are those which result in additional cross-linking, hardening the rubber – and often characterized by a hard, degraded, skin forming on the rubber component. Selection of a suitable rubber and the use of chemical antidegradants can reduce the rate of chemical attack [3].

STRETCHING – A REVERSIBLE PHENOMENON

It is a known fact that elastomers can stretch, but this does not explain why, when the stretching force is removed, the material returns to its original shape. This can be explained by thermodynamics and a much simplified description is given here.

When the rubber is 'at rest' at normal temperature, the chain-like molecules are in a constant state of agitation and are highly twisted due to thermal energy. This is a highly disordered state – described thermodynamically as being a state of high entropy. When the chains are stretched (less twisted), a higher state of order is being imposed, in other words, the chains are being forced into a state of lower entropy. As it is a fundamental law of thermodynamics that entropy strives for a maximum, the driving force is now back towards the disordered state and, as soon as the stretching force is removed, the rubber will contract.

FLUID RESISTANCE

The structure of an elastomer, as mentioned above, comprises a network of chains, meaning that there are gaps between adjacent chains. Indeed, the elasticity of rubber relies on substantial thermal motion of the chains, which would not be possible if the chains were closely packed. The free volume available in the rubber means that some liquids can enter the rubber and cause swelling – sometimes by very large amounts. For example, the ability of oil to swell natural rubber is well known.

A given liquid will have a specific solubility in any given rubber, and knowledge of this will enable designers to avoid excessive swelling when a rubber comes into contact with fluids whilst in service.

Solubility parameters were developed by Charles Hansen as a way of predicting if one material will dissolve in another and form a solution. They are based on the idea that 'like dissolves like' where one molecule is defined as being 'like' another if their chemical bonding is similar. The Hildebrand solubility parameter (δ) provides a numerical estimate of the degree of interaction between materials and can be a good indication of solubility, particularly for non-polar materials such as many polymers. Materials with similar values of solubility parameters are likely to be miscible [4].

INCOMPRESSIBILITY

Another property of rubbers which distinguishes them from other solid materials is their incompressibility. For most practical purposes, other than use under very high pressures, elastomers do not change their volume significantly when deformed. A rubber band may stretch by 600%, but if its volume were measured in the stretched state it would be found to be almost identical to its unstretched volume.

This has important implications for designing with rubbers, as the stiffness of components can be controlled, not just by altering the stiffness of the rubber itself, but by various geometrical designs. This phenomenon, known as the shape factor effect, is described in more detail in standard textbooks and leads to great versatility in design. In particular, it enables engineering rubber components as well as rubber bonded metal seals to be designed with different and controlled stiffnesses.

'O' RINGS AND OIL SEALS

'O' rings

In all industrial hydraulic and fluid systems, pumps, pistons and pipe connections, there exists the requirement of sealing against leakage of the operating fluids or fuels. Toroidal (thick solid ring with circular cross-section) rubber 'O' rings, mostly made of nitrile, neoprene or fluoro-elastomers, possess several advantages in these applications. These universally accepted toroidal seals are light and flexible and, under compression, will deform to follow the contours of the component parts to be sealed. 'O' rings are typically used in rotary, reciprocating and oscillating applications.

The principal operation of the 'O' ring can be described as controlled deformation. This deformation occurs because all 'O' rings are manufactured to allow an initial squeeze or deformation of approximately 8–10% of its cross-sectional diameter. This basically means that the 'O' ring is manufactured to be larger than necessary, causing it to roll and distort when pressure is applied. This deforming squeeze flattens the 'O' ring into intimate contact with the sealing surfaces, and provides the sealing action.

No special tools are required for fitting 'O' rings. In general, they are fitted into a rectangular groove machined in one of the two components to be sealed and the sealing effect is obtained by squeezing the 'O' ring between the bottom of the groove and opposing face. The size of the groove is designed to provide the correct amount of interference (i.e. press fit or tight fit). 'O' rings provide effective sealing in both directions with constant or varying pressure, high vacuum and extreme temperatures [5].

Static application

For static applications, the groove should be designed to allow the 'O' ring to be in contact on all its four sides when in position (Figure 2.3).

Reciprocating applications

In a reciprocating application, the 'O' ring should make contact with the bottom of the groove and the opposing surface. Clearance should be allowed on each side to permit the 'O'ring to move to either side of the groove according to the direction of the pressure (Figure 2.4).

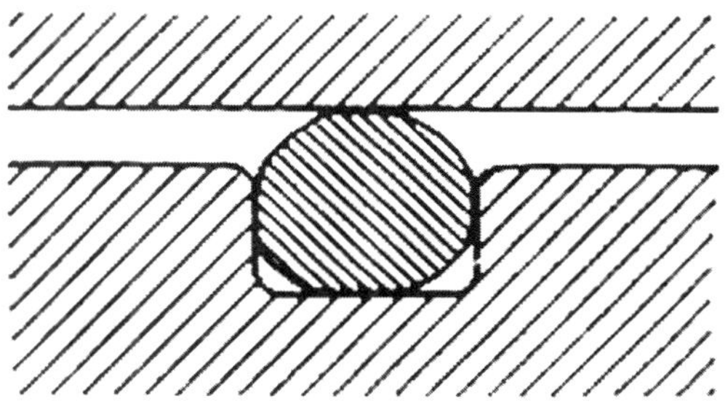

Figure 2.3: Static seal

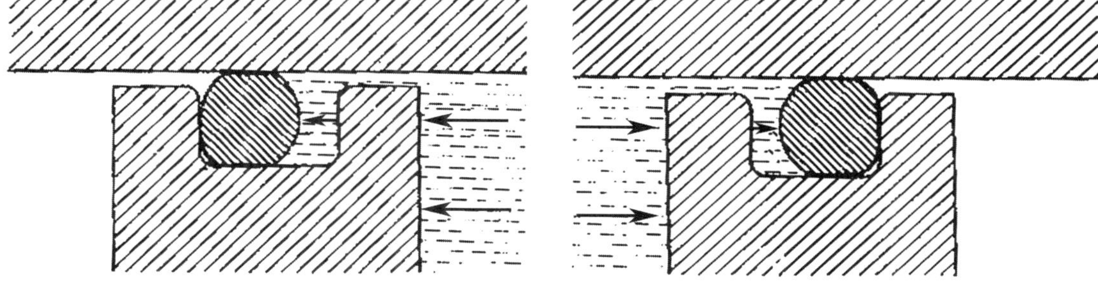

Figure 2.4: Reciprocating application of an 'O' ring

When providing space for an 'O' ring, the annular groove must be large enough to contain its total volume when subjected to diametrical pressures. If the groove cavity is too small, the 'O' ring will be distorted, tend to extrude along the clearances and be mutilated. Any sharp edges on the 'O' ring groove that are likely to cut the ring should be rounded off.

At pressures greater than 2000 psi, the 'O' ring may extrude along the clearances unless an anti-extrusion ring made of a stiff rubber is placed alongside it . The effect of high pressures and anti-extrusion rings is depicted in Figure 2.5.

Cross-section of 'O' rings

To give an example, if an 'O' ring is required to seal between a piston and cylinder of diameters respectively 3.5 inches (9 cm) and 4.00 (10 cm) inches, the nominal cross-section of the ring is 0.25 inch (0.635 cm). But the actual cross-section of the ring should be 0.275 inch (0.698 cm), the difference of 0.025 inch (0.063 cm) being allowed for compression as depicted in Figure 2.6.

It should be noted that the actual cross-section of an 'O' ring is not its nominal cross-section. Most national and international standard specifications have to be referred to for the actual and nominal cross-sectional diameters of 'O' rings.

Applications involving the static, dynamic and reciprocating applications of 'O' rings are depicted in Figure 2.7.

In hydraulic systems, where more than one 'O' ring is used in an application, the following installation requirements should be complied with:

1. Venting should be provided in the space between 'O' rings to prevent fluid or lubricant become trapped between them.

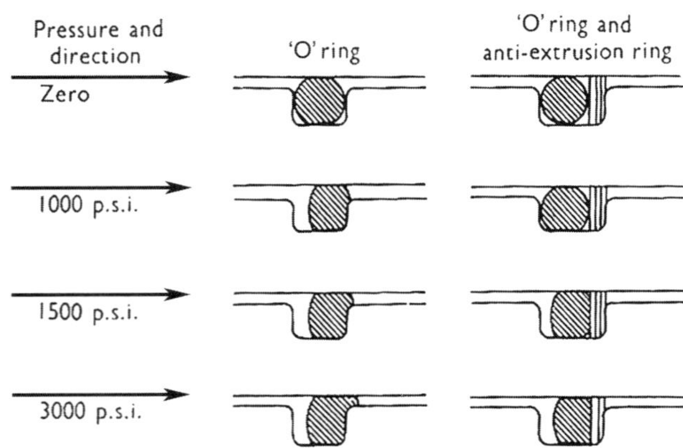

Figure 2.5: Anti-extrusion rings

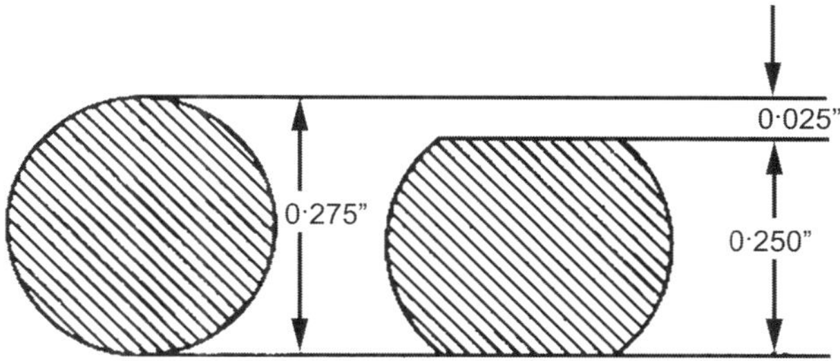

Figure 2.6: Compression allowance of an 'O' ring

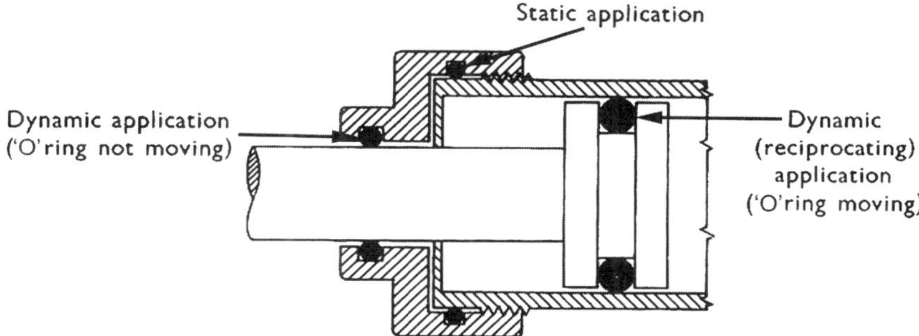

Figure 2.7: Static and dynamic 'O' rings

2. Provision should be made for lubricating 'O' rings not in contact with the fluid being sealed, by using oil-soaked felt rings on each side of the unlubricated ring.

3. Each 'O' ring should be in a separate groove.

Anti-extrusion rings are recommended to be used when:

1. Clearance is greater than normal.

2. An 'O' ring of a particularly soft rubber compound is used.

3. Extra lubrication is used.

4. On all dynamic applications where pressure exceeds 1500 psi.

Where pressure on the 'O' ring is being applied at one side only, then only one extrusion ring is necessary. If pressure is applied from both sides, then two anti-extrusion rings are fitted, one on each side of the ring.

'O' rings for rotary sealing application

When a rubber band is stretched to about 200–300% elongation (Figure 2.8), it contracts if it is warmed and lifts the weight by about 0.25 inch (0.063 cm). The modulus of elasticity or the ability to carry a load increases with temperature. Rubber under constant strain exerts

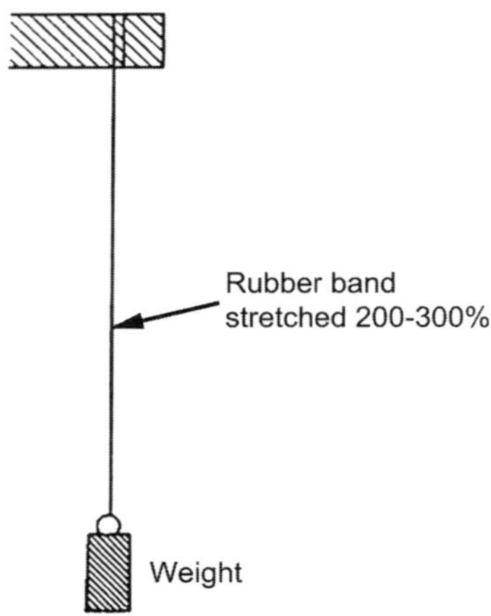

Figure 2.8: Rubber in tension

greater stress. While using an 'O' ring to seal a rotating shaft, an 'O' ring of smaller diameter than that of the shaft is chosen so that it will snap on and make a perfect seal. The seal is thus in a stretched condition, resulting in friction of the rubber against the rotating shaft, so generating heat. This heat in turn causes a stretched 'O' ring to contract since the Joule effect, described above, comes into play.

Gough, in 1805, established that raw rubber becomes warm on being stretched and contracts on being heated. Page, in 1847, and Joule, in 1857, independently reported similar observations. Joule's observations were more quantitative in nature and the aforesaid effect is hence referred to as Joule's effect [6].

This effect in rubber, either raw or vulcanized, contrasts with metals which cool when extended. Change in observed temperature of raw rubber on stretching is an important factor, the study of which is quite useful in the assessment of seal function. The change in temperature of raw rubber on stretching, and the heat of elongation calculated as calories/gram of rubber are given in Table 2.1 from a study by Stafford [7].

A negligible cooling effect was noticed at low extensions (30%). It was noticed that the amount of heat generated increased sharply above 70% extension. The proper Joule effect, according to Vogt [6], is the change of force at constant length or change of length at constant force with respect to temperature, and is reversible. Reciprocating and rotary stresses in seals continuously convert heat into mechanical work. It has been pointed out by Hock that if

TABLE 2.1   Change in temperature of raw rubber on stretching

Percentage elongation	–	30	59	91	146	262	300
Change in temperature (°C)	–	−0.002	+0.020	+0.059	+0.139	+0.657	+0.750

mechanical work done on rubber when it is stretched is calculated in terms of heat, it is only an insignificant fraction (about 2%) of that associated with the stretched rubber. Hence, the heat of elongation, or the Joule's heat as it is called, is something more than the friction heat of rubber under a stressed condition in a seal. Hock's experimental results for heat of elongation are given in Table 2.2.

Despite developments in instrumentation, accurate determinations of heat lost by stretching rubber, or absorbed by allowing stretched rubber to relax, are not yet possible. However, the effect of warming and cooling of stretched and retracted rubber strips, respectively, can be felt by simple observations. For example, if a raw or vulcanized rubber strip at room temperature is stretched rapidly and then is touched by the lip or tongue, one can feel a sensation of warmth indicating that the temperature of the stretched rubber has risen. Similarly, stretch the strip of rubber to a high elongation and hold like that for a few minutes until the heat generated has been dissipated. If the strip is now allowed to retract quickly, one can feel that the strip is cool by touch of the lips or tongue.

The contraction or retraction of a rubber 'O' ring fitted on to a shaft causes a higher unit loading against the shaft due to the Joule effect described above. Thus, the cycle of friction, heating and contraction (stress decay or relaxation) of the rubber ring is repeated until it fails. This can occur within a few minutes at shaft speeds above 200 ft/min. The presence of pressure within the sealing system will cause failure even at slower speeds, since this increases the pressure of the 'O' ring on the shaft. Since rubber is a poor conductor of heat, the surface in contact with the rotating shaft will char before the rest of the ring is affected to any extent. If the assembly is allowed to run for more than a few minutes in this condition, the shaft will become scorched and blue colored from the excessive heat developed. Subsequently, cracks will appear in the ring as it deteriorates, and the fluid being sealed will leak. The remedy followed by many seal designers is to specify the internal diameter of the seal to be slightly more than the shaft diameter by at least 3–5%. The oversized seal does not have the same tendency to heat up. During operation, the temperature levels off after a moderate rise.

A useful case study has been reported by Tatham [8], in which a shaft driven at 3450 rpm was enclosed in a cylinder into which oil was pumped under pressure. One end of the cylinder was sealed with a rubber diaphragm, and the other end with a rubber 'O' ring. The test shaft had a diameter of 0.998 inch (2.53 cm). An undersize ring with internal diameter (ID) of 0.984 inch (2.49 cm) failed in four minutes. During that time, the oil temperature rose from 60°F (16°C) to 295°F (146°C). An oversize ring with ID 1.044 inch (2.65 cm) was run for 100 h with no

TABLE 2.2 Change in heat of elongation of raw rubber on stretching

Percentage elongation	0	186	361	428	621	658	821
Heat of elongation (cal/g)	0	0.84	2.33	2.86	4.58	5.24	6.80

leakage. Oil temperature rose to 152°F (67°C) and leveled off. Under similar conditions seals have been running for 4000 h without trouble.

The compression to apply to rotary sealing 'O' rings has to be determined by experiment. More than real touch contact is needed to offset the compression set of the rubber and any shaft eccentricity. In general, a cross-sectional compression of 2 to 5% is satisfactory for most rotary sealing 'O' ring applications. Surface finish of the shaft is vital in obtaining satisfactory sealing performance. A ground metal finish from 5 to 25 microns as a general engineering practice gives very satisfactory results. If the shaft is too smooth, the seal will not be lubricated properly and so will become over-heated due to friction. If the surface is too rough the seal will wear rapidly. It is desirable to give a chamfering edge to the shaft end over which the 'O' ring will have to pass during assembly so that the ring should not be cut by sharp edges. The groove depths of the rotary 'O' rings should be the same or slightly less than the 'O' ring cross-section. Lubrication of the 'O' ring before assembly is essential.

Rotary sealing rubber compounds, as opposed to static 'O' ring compounds, should most preferably have 80 shore A to 90 shore A hardness. Soft rubber compounds, if used for manufacture of these seals, will tend to drag the rotating shaft, wear out and be heated up, causing failure.

Design aspects of 'O' rings, and other types of seal, that are related to the surface finish of shafts are dealt with exhaustively in industrial documentation, such as in the web pages of various companies [9].

Oil seals

A standard oil seal consists of an outer circular metal disk with an inner flexible rubber which is affixed to the metal during vulcanization. The bonded seal has no loose parts to allow leakage of oil or ingress of any contaminants. This kind of seal is more accurate, and can easily be fitted into a smaller space. An example is shown in Figure 2.9.

The spring shown in the figure is known as a garter spring, and it maintains tension on the sealing lip of the seal. Garter springs are closed coil springs used in the form of a ring, the ends of which are connected together as shown in Figure 2.10.

The depth of the bonded seal can be less and the space between the bore and the outside diameter can be changed for ease of fitting. The bonding of rubber to metal is an important

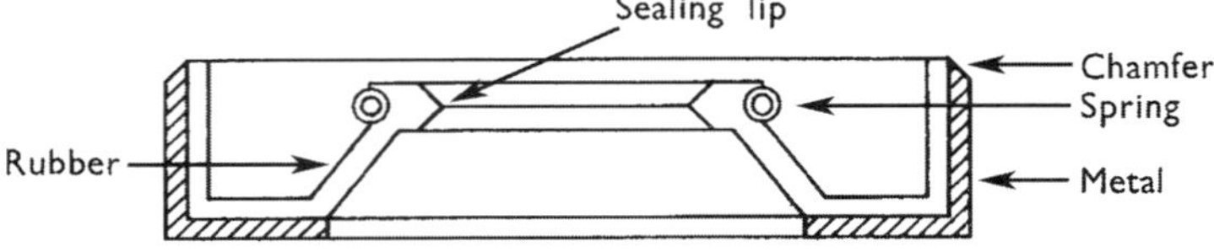

Figure 2.9: Metal to rubber bonded seal

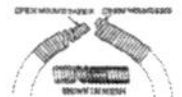

Figure 2.10: Garter spring

factor to be considered in the manufacture of such seals and should be considered carefully, since failure of the bond will cause the seal to fail. The metal case of the oil seal is usually made from mild steel of deep drawing quality which enables blanking, punching, stamping of the steel to the required dimensions.

The edge of the metal is finely ground after seal manufacture in a centerless grinder to enable an interference fit in the oil seal housing. A slight chamfer on the outer diameter (OD) of the seal is desirable for easy assembly. The sealing lip is prepared by buffing, grinding or cutting away the rubber flash which occurs at the sealing edge. A fine sealing edge creates sufficient pressure on the shaft to minimise spring load, leading to lower friction whilst maintaining effective seal performance. The garter spring plays an important role in the efficiency of the oil seal. If its tension is too high, heat will be generated between the sealing lip and the shaft, and result in rapid wear of the lip. If too low, the spring will be ineffective and the sealing lip will be worn away leading to leakage of the fluid.

Another type of seal design has the metal encased in rubber (Figure 2.11).

In this kind of seal, wider tolerances are possible between the OD of the seal and the seal housing. Irregularities of the housing surface can be taken up by the resilient rubber layer on the outside of the seal. However, the rubber covered seal can be blown out under high pressure in a reciprocating application whereas, with a metal case, there is no such danger.

No single physical property of rubbers is responsible for the successful performance of an oil seal or 'O' ring. The ultimate tensile strength, breaking elongation, modulus, shore hardness, creep and stress relaxation in tension and compression loads are all important physical properties that characterize a seal or 'O' ring. Compression strength and set together with stress relaxation or decay are important for effective sealing. The difference in these properties in a swollen seal is highly critical. An optimum swelling value in a fluid medium is a desirable feature. De-swelling decreases the seal pressure against the wall of the housing where the seal is fixed, leading to leakage. Over swelling minimizes the physical properties of the rubber. Seals made of polysulfide rubbers have extreme fuel resistance but undesirably

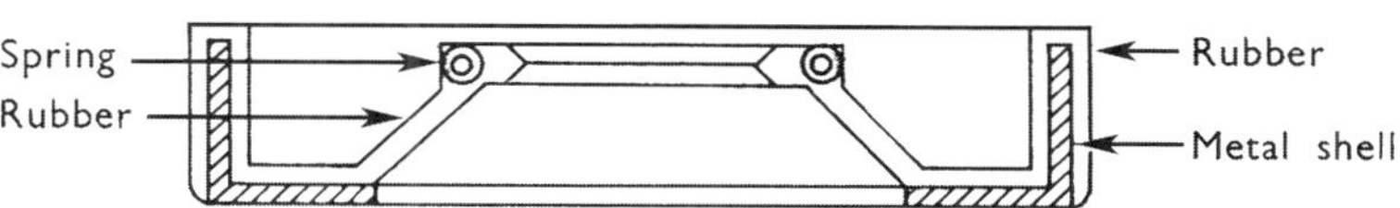

Figure 2.11: Rubber enclosed metal seal

high compression set. The effect of temperature on the seal is an important factor. Swelling under stress can increase at higher temperatures and a suitable compounding technique should be adopted to reduce this effect.

SEALING LIP DESIGN

A sealing lip is in contact with the shaft. It is therefore important that the surface of the shaft be smooth and free from debris before installation. Most lip seals have a pressed fit. Figure 2.12 gives examples of seal lip designs.

MECHANICAL SEALS

These seals operate in a different way to standard oil seals. They are used axially. A seal nose of bronze, carbon or Teflon is mated to a much harder material (i.e. a hardness of 500 brinell) and held in position by a spring. The harder material can be a tightly fitted collar inserted on the shaft or it can be a part of the shaft. The mating surfaces must be true or at high speeds, or the surfaces may break contact and leakage could occur. A mechanical seal and its application are given in Figures 2.13 and 2.14.

These seals are suitable for use at high speeds, temperature and pressures, for instance 5000 ft per minute and 4000 psi pressure at 250°C. Wear on the sealing nose is negligible

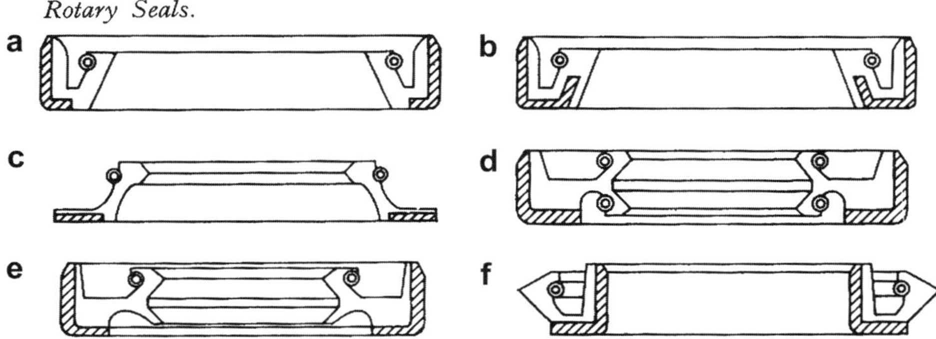

Figure 2.12: (a) Rotary seals having thicker sealing lips are not suitable for higher pressures. They can be used only up to 20 psi; they are not suitable for higher speeds. (b) The seal has an even thicker sealing member with steel reinforcement. This can be suitable for fluid pressures up to 40 psi and is not suitable for high speeds. (c) An alternative seal design where it is difficult to provide conventional housing. This can be fastened to a casing by bolts through the flange. (d) A design suitable for rotating applications where two different fluids have to be kept apart and there is no room for two separate seals; a two in one seal. (e) For applications where dust and grit need to be excluded. (f) A seal that rotates with the shaft, the lip being in contact with the housing

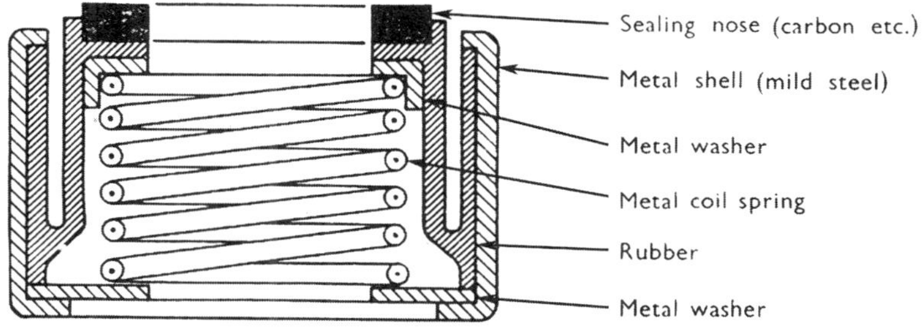

Figure 2.13: Mechanical seal

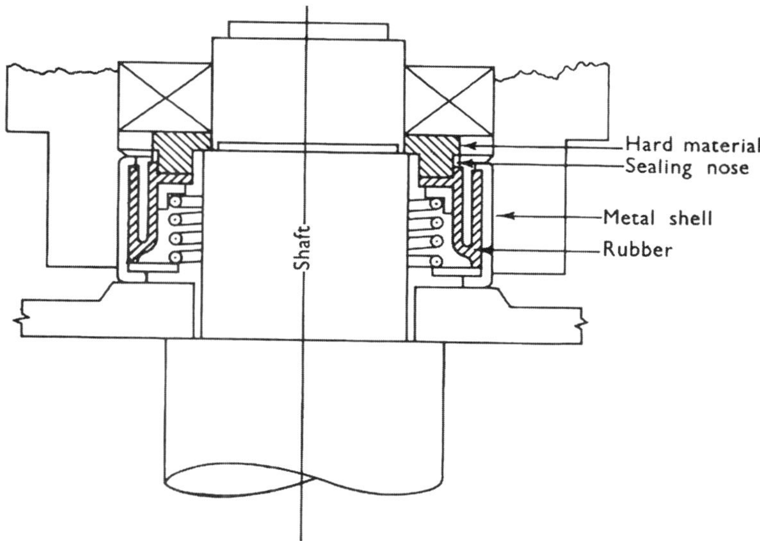

Figure 2.14: The operation of a mechanical seal

and taken by the rubber moving axially under pressure of the coil spring. A mechanical seal eliminates excess leakage but a small amount is acceptable, as it lubricates the seal. This is accomplished by machining the faces that are used in the sealing action to a very smooth finish. These seals are primarily used for rotary sealing applications.

REFERENCES

1 Lebras, L. Rubber – fundamentals of its science and technology, p. 95. Chemical Publishing Co, Inc, New York, 1957.
2 Gere, J.M. Mechanics of materials, 5th edn, p. 22. Brookes/Cole.
3 http://www.merl-ltd.co.uk/index.shtml

4 http://en.wikipedia.org/wiki/Hansen_solubility_parameter.

5 Tatham, N. Oil seals and 'O' rings. Transactions and Proceedings of the Institution of the Rubber Industry, 39, p. T136, 1963.

6 Vogt, W.W. Physics of vulcanized rubber. In: The chemistry and technology of rubber, (David, C.C., Blake, J.T, eds), p. 358. Reinhold Publishing Corporation, New York, 1937.

7 Stafford, G. Physical properties of raw rubber. In: The chemistry and technology of rubber, (David, C.C., Blake, J.T, eds), p. 74. Reinhold Publishing Corporation, New York, 1937.

8 Tatham, N. IRI 39, T149, 1963.

9 http://www.rlhudson.com

CHAPTER 3

SEALS FOR RADIOACTIVE MEDIA – IN NUCLEAR PLANTS

Before going into the subject of radiation-resistant rubber seals and 'O' rings for nuclear plants, the units of radiation, its dosages and activity source will be considered. Since high energy radiation is utilized in the vulcanization process, when sulfur is absent, a basic understanding of nuclear energy, and terminology used in the nuclear industry will be useful.

RADIATION UNITS

Knowledge of these units helps to understand the effects on rubber and humans of radioactive substances such as those that may leak from nuclear reactors.

The 'roentgen' is a unit of measurement for ionizing radiation. The roentgen is the amount of radiation produced by an x-ray tube which, when all the secondary electrons are removed, liberates all positive and negative charges contained in one cubic centimeter of air at 760 mmHg and 18°C (i.e. standard temperature and pressure) so that one electrostatic unit of charge (3.3359×10^{-10} amperes) is measured at saturation current.

One gray is the absorption of one joule of radiation energy by one kilogram of matter. The gray was defined in 1975 in honour of Louis Harold Gray (1905–1965), who used a similar concept, 'that amount of neutron radiation which produces an increment of energy in unit volume of tissue equal to the increment of energy produced in unit volume of water by one roentgen of radiation', in 1940.

The gray measures the deposited energy of radiation. Biological effects vary by the type and energy of the radiation, and the organism and tissues involved. A whole-body exposure to 5 or more grays of high-energy radiation at one time usually leads to death within 14 days. This dosage represents 375 joules for a 75 kg adult. Since grays are such large amounts of radiation, medical use of radiation is typically measured in milligrays (mGy). The average radiation dose from an abdominal x-ray is 1.4 mGy, that from an abdominal CT scan is 8.0 mGy, that from a pelvic CT scan is 25 mGy and that from a selective spiral CT scan of the abdomen and the pelvis is 30 mGy. One gray is equivalent to 100 rad. SI multiples for gray (Gy) are given in Table 3.1.

Rubber Seals for Fluid and Hydraulic Systems 2010; ISBN: 9780815520757

TABLE 3.1 SI multiples for gray (Gy)

Submultiples			Multiples		
Value	Symbol	Name	Value	Symbol	Name
10^{-1} Gy	dGy	decigray	10^{1} Gy	daGy	decagray
10^{-2} Gy	cGy	centigray	10^{2} Gy	hGy	hectogray
10^{-3} Gy	**mGy**	**milligray**	10^{3} Gy	**kGy**	**kilogray**
10^{-6} Gy	**µGy**	**microgray**	10^{6} Gy	MGy	megagray
10^{-9} Gy	nGy	nanogray	10^{9} Gy	GGy	gigagray
10^{-12} Gy	pGy	picogray	10^{12} Gy	TGy	teragray
10^{-15} Gy	fGy	femtogray	10^{15} Gy	PGy	petagray
10^{-18} Gy	aGy	attogray	10^{18} Gy	EGy	exagray
10^{-21} Gy	zGy	zeptogray	10^{21} Gy	ZGy	zettagray
10^{-24} Gy	yGy	yoctogray	10^{24} Gy	YGy	yottagray

Common multiples are in bold face

Gama irradiation dosage units

10 kiloroentgen – 1 kGrays

30 kiloroentgen – 3 kGrays

50 kiloroentgen – 5 kGrays

70 kiloroentgen – 7 kGrays

90 kiloroentgen – 9 kGrays

120 kiloroentgen – 12 kGrays.

Because the energies of the particles emitted during radioactive processes are extremely high, they all fall into the class of ionizing radiation [1]. The practical threshold for radiation risk is that of ionization of tissue. Since the ionization energy of a hydrogen atom is 13.6 eV, a level of around 10 eV is an approximate threshold. The energies associated with nuclear radiation are many orders of magnitude above this threshold value, and fall in the Mev range.

The rad is a unit of absorbed radiation dose in terms of energy actually deposited in the tissue. The rad is defined as an absorbed dose of 0.01 joules of energy per kilogram of tissue. The rem (radiation equivalent man) is the unit of human exposure and is a dose equivalent (DE). The international or SI unit for human exposure is the sievert, which is defined as equal to 100 rem [2]. It takes into account the biological effective of different types of radiation. The biologically effective dose in rems, is the radiation dose in rads multiplied by a quality factor, which is an assessment of the effectiveness of that particular type and energy of radiation.

For alpha particles, the relative biological effectiveness (rbe) may be as high as 20, so that one rad is equivalent to 20 rem. However, for x-rays and gama rays, the rbe is taken as one so that the rad and rem are equivalent for those radiation sources.

Activity of a radioactive source

The curie is the old standard unit for measuring the activity of a given radioactive source. It is equivalent to the activity of one gram of radium. It is defined by:

One curie equals the amount of material that will produce 3.7×10^{10} nuclear decays per second.

The modern, SI unit for radioactivity is the becquerel (symbol B_q), which is defined as the amount of material which will produce 1 nuclear decay per second.

Therefore 1 curie equals 3.7×10^{10} becquerels.

RUBBER SEAL FAILURES IN NUCLEAR PLANTS

Not all nuclear accidents are caused by fluid seal failures. However, an article by Stuart Diamond published in September 16, 1986 and reproduced in *The New York Times* on December 23, 2008, titled 'Nuclear critics charge reactors' and 'O' ring seals pose a threat to safety' is worth reading [3]. An excerpt on 'O' ring failures in nuclear plants is given below:

Thousands of 'O' rings are used as seals in nuclear power plants. An accident could occur, or that a serious accident with some other cause could be worsened by an 'O' ring failure. 'O' rings are precisely engineered versions of the washers that prevent the fluid flow regulators (faucets) from leaking, the gaskets that seal fluids in car engines or the rubber rings that seal jars of preserves. They are used chiefly to isolate materials – whether oil, water or gases – inside parts that are connected and loosened from time to time. In a space shuttle, 'O' rings were designed to prevent gases from escaping from the solid-fuel booster rockets. But cold temperatures apparently makes the seals lose their flexibility. Here, in these applications, rubber compounds having high 'low temperature flexibility' are to be designed for the manufacture of 'O' rings.

In nuclear plants, 'O' rings are used to separate fluids in pumps, valves and switches. They also are part of the system designed to prevent radioactive materials from escaping in an accident, serving as seals for many dozens of hatches, wires or other equipment penetrating the concrete and steel reactor containment building.

A report financed by Public Citizen, a group critical of nuclear power, lists dozens of instances of nuclear plant 'O' ring failures from 1975. For example, faulty 'O' rings at the LaSalle nuclear plant in Seneca, Ill., were blamed for the failure of an automatic shutdown system and an emergency cooling system, the report said.

In August 1985, both main emergency cooling pumps at the Beaver Valley 1 reactor in Shipping port, Pa., were declared inoperable after their 'O' rings were found leaking, the report said.

In various plants, 'O' rings leaking oil impaired the function of shock absorbers, or snubbers, that protect pipes and other equipment in the event of earthquakes. 'O' ring problems were also found to affect diesel generators used for backup power in emergencies.

According to Robert D. Pollard, a nuclear safety engineer at the Union of Concerned Scientists in Washington, that common industrial 'O' rings cannot withstand the high temperatures and radiation of some severe nuclear accidents. Loss of an 'O' ring may not result in the loss of the nuclear plant. But many 'O' rings in nuclear plants are typically qualified to withstand temperatures of only 500 degrees Fahrenheit [260 degrees Celsius], less than half the possible temperature inside a containment should nuclear fuel melt. He says further that high radiation could impair 'O' ring seals and thus undermine emergency systems. Victor Stello, the Nuclear Regulatory Commission's executive director for operations and its top staff member, says that, in a nuclear plant, with thousands of 'O' rings, you will see a lot of failures of 'O' rings. He further says there is nowhere in a nuclear plant where any collection of 'O' rings can cause a major safety problem. He did say, however, that regulators consider 'O' ring leakage 'a significant problem' in the case of certain Westinghouse pumps that circulate cooling water to the reactor. If all alternating current electricity were lost, he said, the four pumps could heat up to the point where the 'O' rings degrade, with a resulting leak of 80 gallons of cooling water a minute. But he said the loss of power to flood the reactor would have to continue for 16 hours before the reactor core was in danger of melting.

Scientists, critical of the nuclear industries' safety record, say that the seals have been failing in nuclear plants hundreds of times during the last decade. As yet none of the failures has caused any very serious accidents, but accidents could well be worsened by an 'O' ring failure.

RADIATION-RESISTANT RUBBER SEALS

Any information related to leakage of radioactively contaminated fluids that results from rubber seal failure, and the effects and dosage levels on humans is useful and should be considered when designing a radiation-resistant rubber 'O'ring or seal. In order to extend the life of rubber seals or 'O' rings in a radioactive environment, hence reducing waste, and the amount of radiation to which the operators are exposed, it is necessary to develop and use highly radiation-resistant rubbers. Many rubber seals manufacturers have done this.

In order to develop suitable radiation-resistant rubbers, irradiation tests have been conducted by seal manufacturers. M/sHayakawa Rubber Co; Ltd in Japan have successfully developed radiation-resistant rubbers, in conjunction with the Japan Atomic Energy Agency (JAEA). They have conducted irradiation tests on various rubbers. Hardness and elongation change results from their trade literature are given in Figures 3.1 and 3.2. 100 G is a grade of rubber from Hayakawa Rubber Co. Ltd. The test methods followed were hardness as per JIS K6253, tensile stress as per JIS K6251, compression set as per JIS K6262 and heat aging as per JIS K6257. Following their study, permitted radiation levels for various rubbers, such as ethylene–propylene (EP), chloroprene (CR), nitrile, fluoro and silicone rubbers are set to

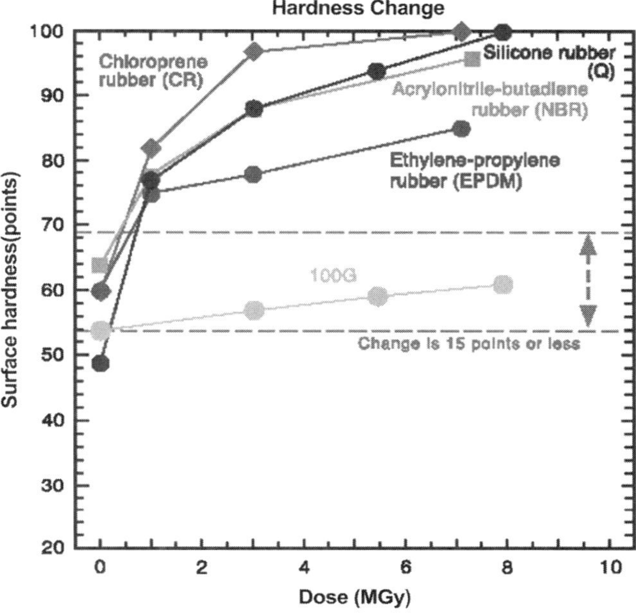

Figure 3.1: Hardness changes at various radiation dosage levels

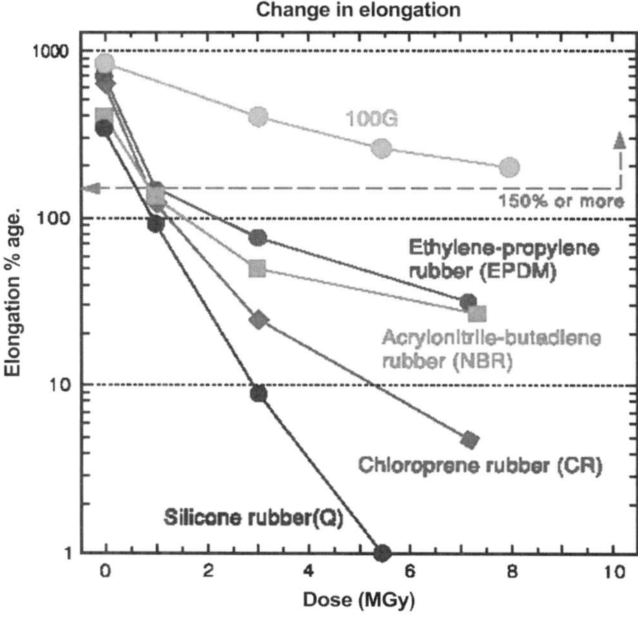

Figure 3.2: Elongation changes at various radiation dosage levels

1.2 MGy–20 MGy [4]. Valuable data on change in hardness, compression set at various doses on 'O' rings and seals made from these rubbers can be found in the report on their website.

CHEMICAL MECHANISM OF CROSS-LINKING BY RADIATION

In the cross-linking of rubbers, free radicals are formed in the molecular chains that form the rubber or polymer. All the chemical and physical processes which cause radical formation in the rubber will cause cross-linking/vulcanization. High energy irradiation is one of the means by which rubbers can be cross-linked through formation of free radicals in the rubber molecular chains. It may therefore be regarded as a kind of indirect vulcanization or cross-linking without a vulcanizing agent. For technical reasons cobalt ^{60}Co is used as a radiation source for vulcanization of rubbers [5]. This is the least powerful radiation source. It emits beta and gamma rays with an average potential radiation output of about 60–140 W, i.e. a gamma ray particle energy of 1.33–1.17 Mev (Mega electron volts) and a beta ray particle energy of 0.306 Mev.

When a high molecular weight polymer, such as natural or synthetic rubber, is exposed to high energy radiation, the free radicals that are produced on the chains react with one another and produce C–C cross-links. Such a linkage, a high stability to heat has compared to the C–S–C links formed in vulcanization, since it contains greater linkage energy, i.e. about 84 kcal/mol. The lowest stability to heat is that of the polysulfide linkage, C–S–S–C, whose weakest link, S–S, has a linkage energy of only about 64 k cal/mol [6]. The bond or linkage energies of vulcanization based on sulfur/accelerator systems, and high energy irradiation, are given in Table 3.2.

It can be seen that cross-links differ from one another in heat stability due to their chemical structure. Those links with least heat resistance (lowest bond energy) are liable to be broken under prolonged heating leading to depolymerization and decay of the rubber. In the manufacture of molded rubber seals using radiation, the following method, as suggested by W. Hofmann, can be employed. The semi-finished rubber blanks are produced by conventional methods, but without vulcanizing agents, accelerators, activators, retarders etc. They are then placed in molds at a temperature of 100–200°C and left for 5–10 minutes, then allowed to cool under pressure. Without being opened, the molds are placed in a radiation chamber. Suitable

TABLE 3.2 Bond energies or the linkage energies of vulcanization based on sulfur/accelerator systems

Type of vulcanization system	Type of linkage	Bond energy (Kcal/mol)
Sulfur/accelerator	–C–Sx–C–	<64
Sulfur/accelerator	–C–S–S–C–	64
Sulfur/accelerator	–C–S–C–	68
High energy irradiation	–C–C–	84

radiation doses vary with composition of the rubber compound. The most suitable radiation dose for vulcanization of nitrile rubbers is about 17 mrep (maximum residual energy path – 1 rep means one roentgen equivalent physical). Polychloroprene rubber requires about 40 mrep, and natural rubber about 90 mrep.

Although oil- and radiation-resistant vulcanized nitrile rubbers have excellent resistance to high temperatures, cross-linking may be accompanied by a splitting reaction and, therefore, the mechanical properties are not always as good as those of conventional vulcanizates. The use of high energy radiation for vulcanization is therefore mainly of academic interest. Different unvulcanized rubbers behave differently in 'irradiation vulcanization'. The irradiation of unvulcanized rubbers gives similar results to irradiating vulcanized rubbers when a splitting reaction takes.

Electricity generated by nuclear power is common in many countries. In Japan, presently 23 percent of total electricity generated depends on nuclear power and this proportion is expected to increase. Although much technological knowledge is required in nuclear power generation, the conversion of heat generated in nuclear reactors into electrical energy by means of a working medium is one of the most important areas of research. Rotary shaft seals made of rubber are important functional components in circulation pumps for the primary coolant. If this sealing system leaks excessively it is likely to result in radioactive contamination along with power failure.

Various types of seals used in nuclear industry are described below.

FABRIC REINFORCED INFLATABLE SEALS

A fabric reinforced seal is a completely molded seal that is built with the same operating principles as a homogeneous inflatable seal. The addition of fabric helps increase the amount of internal pressure under which the seal can operate, where higher pressures are required in a non-enclosed environment (Figure 3.3).

There are several conditions which merit the use of this type of seal, however, the common reason for its use is to resist higher pressures, extremely heavy loads and repeated uniform operation.

Depending on the conditions that the seal will encounter, we can use nearly any fabric available to reinforce an inflatable seal. Some fabrics used will have limitations in the thickness that can be applied within the part, and what temperatures they can withstand during operation. The following are some of the fabrics used:

- Nylon

- Kevlar (fibers of Kevlar consist of long molecular chains produced from PPTA (poly-paraphenylene terephthalamide))

Figure 3.3: Fabric reinforced inflatable seal

- Polyester

- Cotton

- Fiberglass.

Fabric reinforced seals are still prone to over - pressurization and can fail if not properly used. These seals can also fail if they are cut, or if a hole is poked into the side wall of the inflatable cross-section.

In many cases, the use of a proper retention system for a homogeneous seal can compensate for a high pressure environment. With a proper retainer and seal design, a homogeneous seal can often have a thicker profile which can stand up to the most extreme pressures.

Typical application areas for inflatable seals in the nuclear industry

- Pool gates

- Reactor cavities

- Nozzle dams

- Watertight doors

- Butterfly valves

- Airlock doors

- Stud hole plugs

- Canal gates

- Suppression pools

- Motion control devices

As nuclear technology and acceptance of nuclear power has grown, many components suitable for the nuclear industry have become available for the industry. The development of seals for nuclear installations throughout the USA and Canada, as well Korea, India and Argentina, have been made by various rubber-based industries throughout the world.

Inflatable seals are operated through a medium of air or water. Either medium pressurizes the internal chamber in the seal via a valve which, in turn, forces the expansion of the seal profile to its designed shape (Figure 3.4). In addition to air or water, other fluids can be used, such as helium, nitrogen or, in some cases, mineral oils. An inflatable seal can be used in delicate systems.

Inflatable seals are molded in different planes: radially in, radially out, axially and combinations thereof (Figures 3.5–7). They will conform to many different contours. They can be made in strips with closed ends or in continuous loops. The bend radius will vary from seal to seal and the seal can be molded to exact configurations, even of very sharp corners.

Advantages of an inflatable seal

- Provides a leak-proof closure, yet allows clearance when needed

- Simplifies the design of the structure and hardware

- Minimizes the need for close machining and/or small fabrication tolerances

- Not subject to compression set which negates effectiveness of other seals

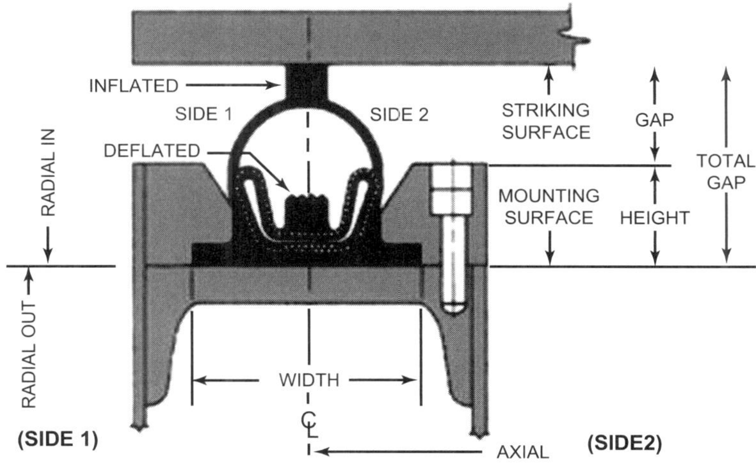

Figure 3.4: Inflatable seal in working condition

Figure 3.5: Contours of inflatable seal

Figure 3.6: Inflatable seal with sharp corner

Figure 3.7: Inflatable seal with bend radii

NON-REINFORCED INFLATABLE SEALS

Non-reinforced inflatable seals are homogenized seals manufactured only from rubber without any reinforcement but having thicker wall sections. These inflatable seals can be used for a variety of applications where sealing two surfaces that move in relation to each other is required. By introducing a medium such as air or fluid, the seal will expand at a specified rate until a determined size is achieved. The rates and sizes of expansion of an inflatable seal are determined by its shape (Figure 3.8). Nearly all inflatable seals, reinforced or non-reinforced, are custom fabricated for a specific application.

INFLATABLE SEAL OPERATION AND CAPABILITIES

Inflatable seals are operated via a medium such as air or water, however other specific gases or liquids can also be used. Where oxygen is not desirable, helium or nitrogen may be substituted. In some cases, mineral oil may be used. Temperature issues can play a significant role in determining which medium to use. For extremely low temperatures, hot liquid or air can be introduced to keep the actual temperature of the rubber compound from falling below its lowest practical operating temperature. Freezer doors are a good example; where hot air may be used to prevent the rubber from freezing and sticking to metal surfaces. Likewise, under condition of high temperature cold air or water can be used to prevent the compound from overheating.

Figure 3.8: Customized non-reinforced inflatable seal

SEALS IN PUMP ASSEMBLIES IN NUCLEAR PLANTS

Pump assemblies are very important equipment in atomic energy installations; they have a considerable effect on the capital and operating costs of atomic electrical stations (AES) [7]. The reliability of nuclear reactor operation is determined by the suitability of the pumping equipment, which depends on the technical characteristics of the components which make up the mobile and immobile junctions. Rubber seals, characterized by such valuable properties as elasticity, corrosion resistance, simplicity of fabrication and comparatively low cost are widely used as construction materials. They ensure reliable hermetization of chambers in pneumatic, hydraulic and vacuum systems at different temperature and pressure conditions. In operation under such conditions as an AES, rubber materials, in addition to the action of such factors as chemical media, high and low temperatures, pressure, and mechanical loads, are also subjected to the destructive action of ionizing radiation. When rubber articles are irradiated, especially when stretched as compressed, they lose their elasticity and hence their ability to act as a seal. The problem of choosing rubber materials able to withstand radiation when designing pump equipment for AES has hence become a challenging one. This is the more important because, compared to metals, ceramics or glass, rubber materials are subjected to considerably higher levels of radiation. The use of radiation-resistant rubbers in AES pump equipment increases its operational reliability and reduces the number of overhauling operations, which are often carried out under conditions of ionizing radiation. The absorbed dose of ionizing radiation at which the material reaches a critical state serves as a quantitative measure of the radiation resistance of a rubber. The radiation resistance of rubbers depends on their composition and the chemical nature of the base rubber. Under the action of ionizing radiation, processes of radiochemical cross-linking and degradation take place in different rubbers in different ways, as a result of which their properties change irreversibly. Rubbers based on butadiene–acrylonitrile butadiene-styrene, ethylene–propylene, butadiene, fluorine-containing or siloxane polymers and so on are materials which predominantly cross-link under irradiation. Their properties change as follows: increase in strength in extension (in the initial stages of irradiation), in hardness, brittleness, elastic moduli in compression or extension, build-up in residual deformation, decrease in relative elongation at break, in relative residual elongation and loss of elasticity. Rubbers based on isoprene–isobutylene (butyl) rubber are degraded on irradiation; in their case a decrease in strength, hardness, elastic moduli in compression or extension and a rise in relative and residual elongation at break, appearance of tackiness, flow and stress relaxation occurs.

Rubbers based on natural rubber, and synthetic rubbers containing protective aromatic groups in their structure (such as butadiene–styrene rubbers) show the greatest resistance to the action of ionizing radiation. In their case, the rates of the cross-linking and degradation processes are close, so over a given range of absorbed doses of radiation, a stability of the moduli in

TABLE 3.3   Chemical resistance of some common rubbers used in nuclear industry

	Tear	High temp	Low temp	Oil	Weather	Acid	Alkali
EPDM	2	2	2	3	1	1	1
Neoprene	2	2	2	3	1	1	1
NBR	2	2	2	1	1	1	1
Silicone	3	1	1	3	1	3	1

EPDM: ethylene–propylene terpolymer; NBR: nitrile butadiene

compression and extension, as well as strength, relative elongation at break, and retention of elasticity is noted.

The chemical resistance of some rubbers used in nuclear plants is given in Table 3.3.

REFERENCES

1 http://hyperphysics.phy-astr.gsu

2 Radiation Safety Manualf:http://www.orcbs.msu.edu/radiation/programs_guidelines/ radmanual/13rm_radiounits.htm).

3 Diamond, S. Nuclear critics charge reactors' 'O' ring seals pose a threat to safety. *The New York Times* December 23, 2008

4 www.hrc.co.jp - Hayakama Rubber Co

5 Hofmann, W. Vulcanization and vulcanizing agents. Farbenfabriken Bayer AG, Leverkusen 1965. English translation Maclaren and Sons Ltd, London, 1967

6 Hofmann, W. Vulcanization and vulcanizing agents, pp. 21, 130. Farbenfabriken Bayer AG, Leverkusen 1965. English translation Maclaren and Sons Ltd, London, 1967

7 Bukanova, N.N., Davydov, I.V., Zavolokova, V.G., Kostereva, G.M., Starkova, N.A. Rubber seals for radioactive media. Translated from I himicheskoe i Neftyanoe Mashinostroenie, 3, 12–13

CHAPTER 4

RUBBER SEALS IN AIRCRAFT

Leakage in fuel and hydraulic fluid systems of aircraft are viewed extremely seriously. Safety is a compelling factor that demands high quality from rubber seals to be used in aircraft components. Hence, aircraft seals are manufactured with extra precision and extremely strict quality control measures for materials and their processing. Strict adherence to the specification of raw materials, processing methods, testing procedures, and dimensional standards are required.

Aircraft components, including those for fluid systems, are designed to minimize weight. However, specifying small grooves for 'O' rings can lend to leakages, which must be viewed seriously. Weight is obviously money and higher fuel consumption but, in order to mend a leak, it is necessary to dismantle the whole assembly of the aero engine, frames or rotors to locate and replace the failed seal leading to unrealistic maintenance schedules and long downtimes for the aircraft. At times, component alterations need to be done because of seal performance, so the dependability and quality of a rubber seal are as important as the components themselves.

The purpose of aircraft fluid seals and 'O' rings is to prevent leakage of hydraulic fluid and thereby preserve the pressure in the hydraulic system. Seals are installed in most of the units that contain moving parts, such as actuating cylinders, valves and pumps. Seals used in hydraulic and pneumatic systems and components are of two general classes, called packings and gaskets. Packing rings are used to provide a seal between two parts of a unit that move in relation to each other, whereas gaskets are used as a seal between two stationary parts. The handling, installation and inspection of rings, seals and gaskets in aircraft hydraulic systems are extremely important, as a scratched or nicked sealing ring can cause the failure of a system unit at a critical time. Typical shapes of hydraulic packing rings, seals and gaskets are shown in Figure 4.1.

The 'O' ring is the most common type of seal used for sealing pistons and rods, has it is effective in both directions. A color code is given on the 'O' ring to identify the type of system in which it is used and the name of the manufacturer. 'O' rings have either a series of colored dots arranged clockwise or a colored stripe around the outer circumference. The first dot clockwise, or the stripe indicates the system in which the 'O' ring is used. The following color codes are generally used: red indicates use in the fuel system, blue indicates use in hydraulic and pneumatic systems, yellow indicates use in oil systems, and a white dot appearing first indicates a non-standard seal.

Rubber Seals for Fluid and Hydraulic Systems 2010; ISBN: 9780815520757

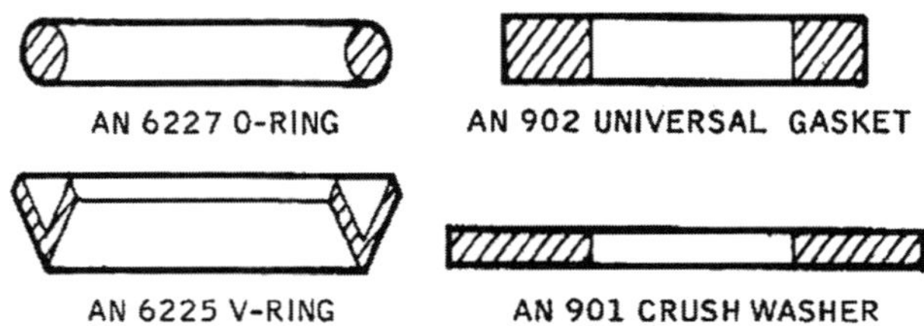

Figure 4.1: Shapes for standard 'O' ring seal, chevron seal, universal gasket and crush washer

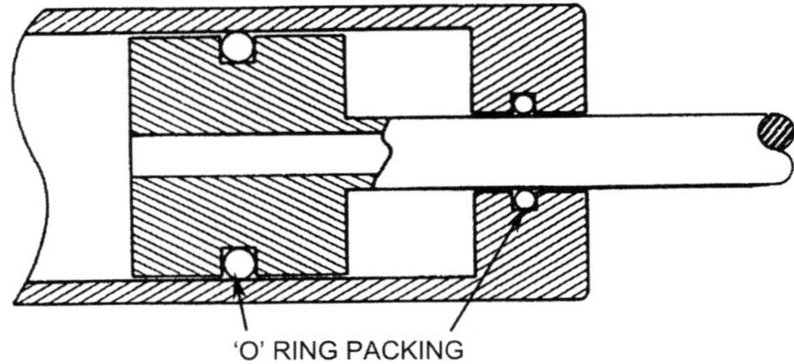

Figure 4.2: Typical 'O' ring installation

Generally, the 'O' ring requires no adjustment when it is installed. Figure 4.2 shows a typical 'O' ring installation. However, a few guidelines should be observed at installation of an airborne seal or early failure may result. Always install a new 'O' ring with each disassembly. Whatever the condition of the ring when disassembling components, a mandatory replacement of the sealing ring is followed in aircraft maintenance and repair schedules. Old seals are not used, even though they may appear to be still serviceable. The ring and the groove where it is installed must be lubricated freely with the same type of fluid that is used in the system, unless otherwise stipulated in the aircraft repair manual. The 'O' ring must not be twisted when it is installed. If the hydraulic pressure is too high, i.e. more than 1500 psi, the ring can become pinched between the moving and stationary parts of the system, as shown in Figure 4.3, thus damaging the ring, leading to seal failure. In such situations, where high pressure is encountered in the system, backup rings are installed in conjunction with 'O' rings. The purpose of the backup rings is to prevent high pressures from destroying the sealing ring by giving additional strength to the ring. Backup rings are usually made of Teflon. Installation of an 'O' ring with backup rings on both directions respectively is shown in Figure 4.4.

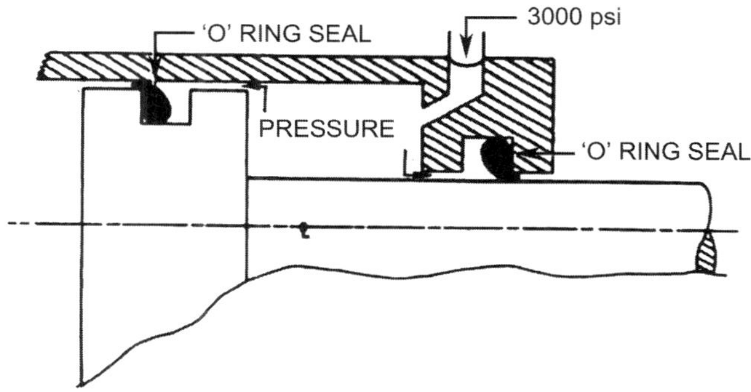

Figure 4.3: 'O' ring in a pinched or extruded condition

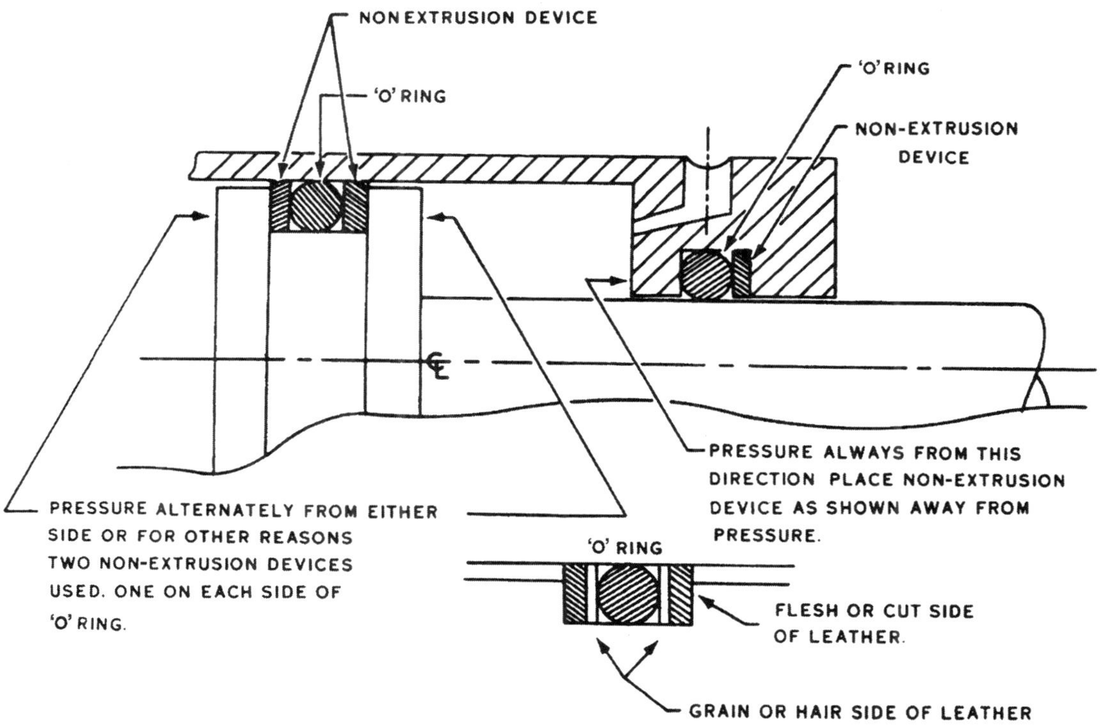

Figure 4.4: 'O' ring installation with backup ring

HYDRAULIC SYSTEMS IN AIRCRAFT

In the consideration of seals or 'O' rings for the aircraft industry, it is worth knowing the basic fluid flow mechanism in the hydraulic system, since the purpose of seals in aircraft is to preserve the pressure in such a system. Every aircraft hydraulic system has two major parts, the power section and the actuating section. The power section provides fluid flow, regulates and limits pressure, and carries fluids to the various selector valves in the system. The actuating section or subsystems contain the various operating units, such as wing flaps, landing gear, brakes, boost systems and steering mechanisms. The power section may either be an open or closed system, using an engine driven pump, or a pump driven by an electric motor. An open system means that there is a free path for fluid flow back to the reservoir, until one of these units is actuated and directs the fluid flow from the reservoir valve. Fluid flow is then directed to the actuators (cylinders and motors) and pressure will build sufficiently to move these units. In closed systems, the pressure developed by the engine-driven or electric motor-driven power pump may be regulated by a pressure regulator, by the power pump, or by a pressure switch. A closed system will always have fluid stored under pressure whenever the pump is operating. An open system has fluid flow, but no appreciable pressure in the system when the actuating mechanisms are idle. A closed system directs fluid flow to the main system manifold and stores fluid under pressure in the portion of the system which leads to all the selector valves. Any seal failure will therefore disturb the flow mechanism in the hydraulic system, leading to undesirable events during flight.

Increased speeds of aircraft lead to higher temperatures within the fluid system. As a result, operating fluids attain higher temperatures which are associated with changes in fluid viscosity and pressures which leads to a higher likelihood of leaks. Present day aircraft operators, whether they be civil or military, fly at higher altitudes where temperatures vary between $-50°C$ and $+200°C$, in atmospheres of low and high concentrations of oxygen and ozone, and with system pressures of more or less than 3000 psi. Hence, airborne rubber seals often encounter extreme temperature variations, reactive fluids, mechanical stress and strain, and friction. This means that a seal needs to have high degree of thermal and oxidative stability, high resistance to fluids, and to dynamic wear.

Downtimes for aircraft maintenance are of great economic concern so if seal life could be increased, a large saving in the ultimate cost of aircraft operation would result.

SEALING MATERIALS

Leather was used as a sealing material in the early days, but now seals made out of rubber are universally accepted. Before the advent of oil resistant, acrylonitrile rubbers natural rubber seals were used that had appropriate compounding to give resistance to oils and fuels. Far reaching compounding developments have been made over the past decades giving highly dependable rubber compounds for fuel resistance up to 200°C. Improvements in seal design and assembly procedures have improved seal performance greatly in aircraft. Associations like the US Society of Automobile Engineers, British Hydromechanics Research Associations and other aircraft operators have contributed by giving a lot of data to rubber manufacturers, which would have been difficult to collect without their co-operation. The advent of newer types of rubbers has contributed towards the increased life of seals due to their inherent properties at extreme conditions. Rubber compounding and manufacturing of products such as these are generally kept as closely guarded secrets, although vast sources of information is available in fluid sealing technology and aircraft maintenance manuals, a text dealing with rubbers in aircraft sealing applications is not available in the technical literature.

Fluororubbers and silicones are the best materials to make seals resistant to extreme temperature conditions. Viton, a fluororubber marketed by M/S Du Pont USA, can operate at 200°C in contact with oils and lubricants. Polysulfide rubbers have low compression set, but exhibit excellent fuel resistance. For improved compression set, it is admixed with relatively cheap conventional nitrile rubbers. The fluorine-containing rubbers, such as fluororubber, possess outstanding resistance to heat, fuels and hydraulic fluids coupled with extremely good aging characteristics. The reversible physical effects with respect to the ultimate tensile strength of this rubber vulcanizate is a noteworthy phenomenon since this is not associated with the deterioration of the rubber itself. Fluororubber is resistant to most fluids used in the aircraft industry, such as synthetic, ester type lubricants, aromatic and aliphatic hydrocarbons, and water.

The high frictional heat developed in aircraft fuel and hydraulic systems increases the temperature of the components which stresses the rubber 'O' rings and seals. Under such conditions, fluororubber is the most promising sealing material. Other fluorine containing rubbers, such as polyfluoroalkoxyalkyl acrylate and chlorotrifluoroethylene vinyledene fluoride, have emerged out of research programs sponsored by USAF, and are noteworthy for their extraordinary inertness towards chemicals, as well as their high oxidative and thermal stability.

DESIGN OF MOLDS AND PARTING LINES

Generally, it is desirable to have single cavity molds in seal production to ensure uniformity between of seals and 'O' rings. Seals made out of multicavity molds are prone to differ in dimensions between cavities. This is so because the volume of rubber in each cavity cannot be controlled with precision during normal manufacturing practice. Also, the quantity of seals required for aircraft does not warrant a multicavity mold. In other words, the total requirements of seals or 'O' rings can easily be met during the time between overhauling or manufacturing of the aircraft component involved. Practice differs between manufacturers of aircraft seals in locating the parting line of an 'O' ring that is formed during molding. Where the parting line can be made invisible or extra smooth by precise cutting of the metal, and accurate registration of the parts of the mold, the same can be provided at the sealing surface of the ring, i.e. at an angle of 180°. However, this depends on the skill of the mold maker. If the mold is expected to be in use for a longtime, and will become prone to overlap, then it is desirable to shift the parting line away from the sealing face by 45°. A smooth parting line at 180° angle and smooth sealing surface with 45° angle, together with an overlapped sealing face, are depicted in Figures 4.5–4.8. A prominent parting line at the sealing surface will increase friction, reduce sealing pressure causing failure, and may sometimes contaminate the

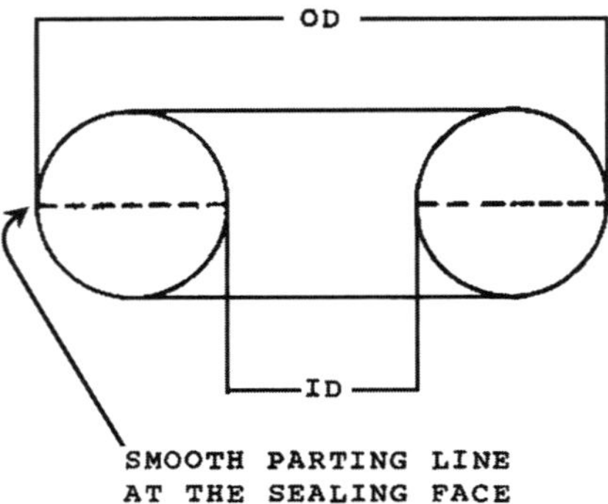

Figure 4.5: Parting lines

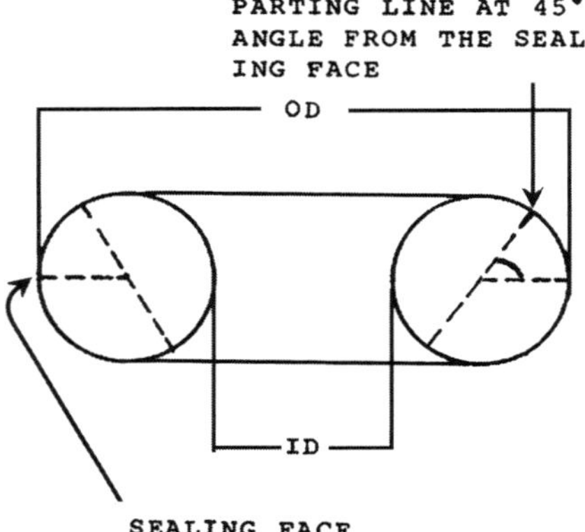

Figure 4.6: Parting lines

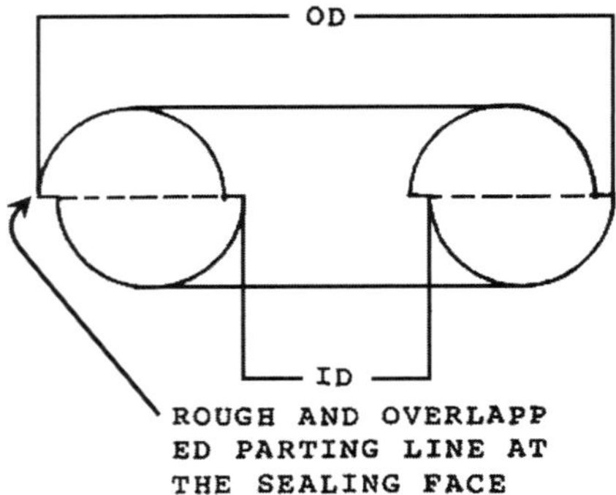

Figure 4.7: Parting lines

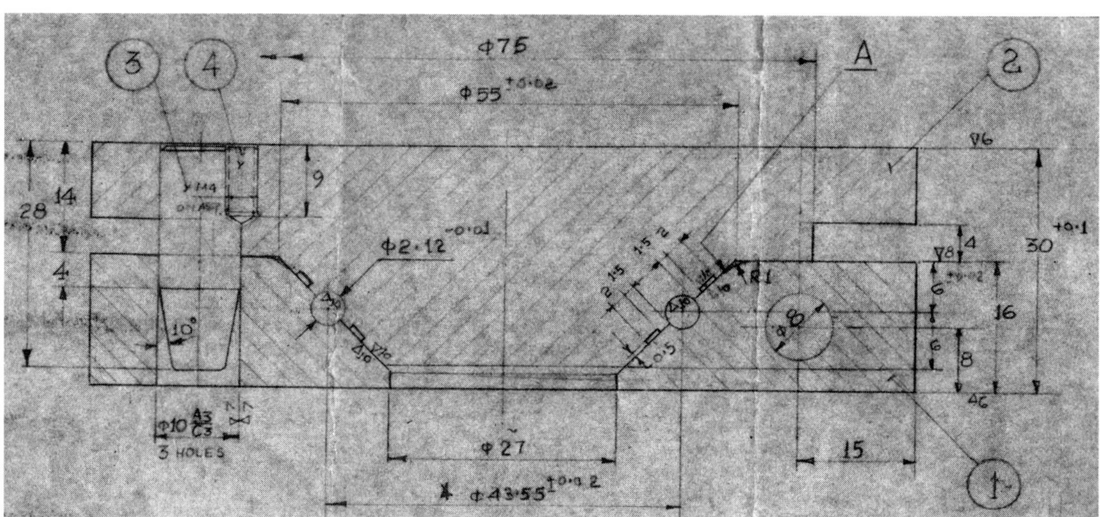

Figure 4.8: Mold design for an 'O' ring with a parting line at 45° angle. 1. Mold; 2. punch; 3. dowel pin; 4. screw holding the pin; A. Mating surfaces. From General Engineering Section Base Repair Depot, Indian Air Force, M/O Defence, Government of India

fluid medium. Repeated checking of dimensions and tolerances is extremely important in aircraft seal manufacture, in order, to ensure precision.

DESIGN OF SEALS FOR AIRCRAFT

The penetration of fluids through a rubber 'O' ring or seal is an important factor to be considered when selecting a suitable base rubber for manufacture. The degree of swelling and the thermal coefficient of the rubber and that of the metal components should also be considered. For example, the coefficient of thermal expansion of rubber is ten times that of steel. The volume of the groove housing the ring should be sufficient to accommodate it when swollen or, alternatively, the rubber should be chosen such that its maximum volume will be less than the volume of the groove. This will prevent the extrusion or pinching of the swollen rubber into the shaft–cylinder interface, thus leading to system failure. The uptake of a fluid by a rubber seal vulcanizate is altered by application of stress. It is increased by stretching, and decreased by compression. Usually compression is the main strain to which the rubber is subjected. Increasing the temperature of the system has a phenomenal effect upon the sealing ring. Firstly, the 'O' rings expand to more than the volume of the groove in which they are installed. Secondly, the increase in temperature will increase the rate of swelling. Thirdly, there is an increase in the cross-link density and scission of the rubber cross-links.

On the other hand, at low temperatures, the rubber tends to shrink and cause leakage. There is a reduction in the cross-sectional diameter of the ring when it is stretched during installation.

With thin walled cylinders, as are generally seen in aircraft components, a diametrical stretch occurs due to high fluid pressures. In the design of rubber 'O' rings or seals, establishing the dimensions of the rings, selection of the base rubber and its compounding with additives should all be considered carefully.

To assess the storage stability and serviceability of rubber seals, accelerated aging tests are often conducted. It is, however, difficult to give quantitative correlations between the results of these tests and the actual aging of the seals in service. The changes in a rubber vulcanizate kept under proper storage conditions are generally physical in nature, caused by slow changes in the rubber molecules, due to scission and cross-linking reactions. Accelerated aging tests, conducted at elevated temperatures, and in the absence of oxygen, provide information regarding a mechanism of aging that is similar to natural aging on the shelf at lower temperature. Although rubber vulcanizates are generally stored under standard conditions, away from the influences of oxygen, ozone, light, chemicals, oils, acids, etc, they are liable to change their physical properties over time due to slow chemical reactions in the cross-linked network. This change is very slow at room temperature. Prescribing a storage life time for a sealing ring in the aircraft industry is important, although it is often disregarded. In such instances, accelerated aging test results can be a useful measure for approximating the shelf-life of seals.

Environmental conditions likely to be encountered by aircraft in coming years can be expected to become more severe, coinciding with advances in aerospace technology. Specialty synthetic rubbers will be needed for the aircraft industry's future sealing requirements. Although rubber compounding techniques on these specialty rubbers, as well as conventional nitriles and neoprenes, have been adopted by seal manufacturers, much research is still needed from leading rubber and chemicals manufacturers' R&D laboratories and institutions.

CHAPTER 5

RUBBER SEALS FOR OIL FIELD SERVICE

Conventional sealing techniques and designs, including those using highly temperature resistant rubbers such as nitrile to fluororubbers, used to be adequate for most applications. In the oil field industry, seals come into contact with fluids such as well fluids, completion fluids and stimulation fluids. This means that seals need to be designed to withstand these harsh conditions, so some introductory information about these fluids and their corrosive characteristics would be useful, since they have an adverse effect on seal function. This is given below.

WELL FLUID

When drilling or completing wells in earth formations, various fluids are typically used. The fluid is often water based. Such fluid is referred to as 'well fluid', and its uses include: lubrication and cooling of drill bit cutting surfaces while drilling; controlling fluid pressure to prevent blowouts; maintaining well stability; suspending solids in the well; minimizing fluid loss into and stabilizing the formation through which the well is being drilled; fracturing the formation in the vicinity of the well; displacing the fluid within the well with another fluid; cleaning the well; testing the well; abandoning the well, or preparing it for abandonment.

COMPLETION FLUID

Completion fluids improve well productivity by reducing damage to the producing zone and they can help prepare, repair, cleanout and complete the well bore during the completion phase. They are typically brines (chlorides, bromides and formates) but, in theory, could be any fluid of proper density and flow characteristics. The fluid should be chemically compatible with the reservoir formation, and is typically filtered to a high degree to avoid introducing solids to the near-wellbore area. Regular drilling fluids are seldom suitable for completion operations due to their solids content, pH and ionic composition. Drill-in fluids can, in some cases, be suitable for both purposes.

Rubber Seals for Fluid and Hydraulic Systems 2010; ISBN: 9780815520757

STIMULATION FLUID

Stimulation fluid is a treatment fluid used to stimulate production in a well. The term is most commonly applied to matrix stimulation fluids. Most matrix stimulation fluids are acid or solvent based, with hydrochloric acid being the most commonly used due to its reaction characteristics and relative ease of control. Matrix stimulation is a process of injecting a fluid into the formation, at pressures below the fracturing pressure, to improve the production or injection flow capacity of a well.

In an oil field environment, the conditions are worsened for 'O' rings or rubber seals because of the following:

1. During the oil field operations, the combination of carbon dioxide and hydrogen sulfide with or without water creates aqueous and non-aqueous electrolytes.

2. High concentrations (or partial pressures) of gas at temperatures above critical can act as super-critical solvents. Rubbers in this environment are subjected to high swells leading to subsequent extraction of plasticizers, low molecular weight polymers etc.

3. Explosive decompression in the oil field due to pressure or temperature shifts can cause catastrophic seal or 'O' ring rupture.

4. The inevitable use of fluids such as hydrochloric and hydrofluoric acids, bases, metallic halides, carbonates, and gases like carbon dioxide and nitrogen, create another set of problems for the seal.

5. Oil field conditions are very harsh and unique and, therefore, there is very little technology transfer potential.

The design of suitable compounds for use in the oil industry is complicated by the following factors:

1. Differences in the types of carbon blacks used in the oil field service lead to seal explosive decompression problems. The molecular weight of the base rubber also is very critical

2. Metal oxides are used in the curing systems of seal compounds. These are soluble in the inorganic acids that are used in the oil exploration industry, and this leads to the rubber swelling considerably.

3. If non-black fillers are used, they produce excessive swelling and softening of 'O' rings in the aqueous medium encountered, because oxides become hydroxides, and silicates are converted to water-soluble bicarbonates by the carbon dioxide and water that is present.

4. Seal degradation occurs when in contact with high pH (basic) and low pH (acidic) fluids in the systems.

These factors show that compounds suitable for use in oil fields are difficult to produce because they have to be based on high molecular weight rubbers and highly reinforcing fillers. Unfortunately, such compounds have undesirable higher compression set values.

A chemist, chemical or corrosion engineer involved in the oil industry, will certainly appreciate the fact that rubber is an engineering material as molded rubber seals or 'O' rings are used under conditions of strain and stress. Compounding for either of these conditions needs considerably different approaches. The following three basic factors need to be considered:

- choice of rubber or elastomer

- choice of fillers

- degree of cross-linking.

Rubber can be considered as a thermodynamic system, with the first law explaining the property of elasticity quantitatively and the third law illustrating the thermal pressure created in the molding process. Rubber is a polymer; its structure is a chain of repeating units. The lengh of the chain is indicated by its Mooney viscosity, which decreases as molecular weight (or chain length) decreases. Low die-swell also indicates low molecular weight. The narrow molecular weight distribution of high molecular weight rubbers gives better molding characteristics. A perfect rubber selection for sealing devices in the corrosive environment of the oil field industry is very difficult due to the dynamic conditions that prevail. This contrasts with the chemical industry, where, although exposed to a corrosive environment rubber is used under static conditions, such as in storage tanks or piping systems handling corrosive chemicals.

Fillers have numerous functions in a rubber compound, but they are all due to the increase in the viscosity of the formulation. For more rigorous oil-field sealing applications, particularly those in the drilling hole, the preferred filler is carbon black. Since oil production environments involve methane (CH_4), hydrogen sulfide (H_2S), and carbon dioxide (CO_2) gases as well as extreme pH values, non-black fillers can create complications. There is a strong possibility that such operating conditions will damage the reinforcement of the rubber.

EXPLOSIVE DECOMPRESSION

The reinforcing effects of fillers cannot be overlooked while specifying the performance requirements of an 'O' ring or a seal for oil field use. Briscoe et al [4], in one of their many papers, gave additional insight into the carbon filler–rubber adhesion during reinforcement. They observed that the degree of adhesion of a filler in a rubber influences the carbon dioxide gas uptake, as indicated in Table 5.1.

Table 5.1 shows that the coated filler absorbs more carbon dioxide, leading to blistering in the seal, known as explosive decompression. It becomes evident that rubber compounds that have low permeation rates are more resistant chemically to permeation of gases and are the best to use to minimize blistering due to cavities in the rubber or polymer matrix.

Table 5.2 [5] shows the effects of four important filler variables on the physical properties of a sealing compound, those being quantity, surface area, structure and surface reactivity. The resistance to blistering, or explosive decompression, is found to increase with an increase of all the four variables. Table 5.2 shows how these properties change as each of these variables increases. It should be studied in conjunction with Figure 5.1, from which the specific properties can be optimized. Since each of these variables can affect properties differently, and cross-link density of the compound will have yet another effect, (shown Figure 5.1), the choice of properties to be optimized should be selected very carefully, and should be limited.

The function of cross-linking requires little discussion in the matter of seal design as it is covered in many texts. The cross-link density directly affects physical properties such as heat build up, tear strength and elongation.

EFFECT OF INCREASING MOLECULAR WEIGHT

In the case of amorphous rubbers, the effect of increased molecular weight on their physical properties deserves careful consideration. Table 5.3 lists many of the properties that improve as the molecular weight increases. The enhancement of rubber properties with increased molecular weight has been known for many years but development has been limited, because of the difficulty of processing these high molecular weight rubbers. However, for high pressure and high temperature sealing applications, as in the oil industry, only high molecular weight rubbers are suitable, since they possess low compression set along with other desirable functional properties.

A careful study of Tables 5.1–5.3 and Figure 5.1 will indicate why a rubber chemist spends a lot of time in designing compounds with various cross-link densities. Tear strength,

TABLE 5.1   Filler/gas uptake relationship

Condition	Percent gas uptake
No filler	133
Untreated filler – 20% by volume	145
Coated filler – 20% by volume	150
Filler silane coated – 29% by volume	120

TABLE 5.2 Effects of filler

	Increasing quantity	Increasing surface area	Increasing structure	Increasing surface reactivity
Hardness	↑	↑	↑	↑
Dynamic modulus	↑	↑	↑	↑
Static modulus	↑	↑	↑	↑
Tensile strength	\	↑	→	→
Elongation	↓	→	↓	\
Compression set	↑	↑	↑	↓
Tear strength	\	↑	→	↑
Fatigue life	\	↓	↑	↑
Abrasion resistance	\	↑	↑	↑
Impact strength	\	↑	↑	↑
Heat build up due to hysteresis	↑	↑	↓	↓
Extrusion resistance	↑	↑	↑	↑
Blister resistance	↑	↑	↑	↑
Electrical conductivity	↑	↑	↑	→
Processing				
Incorporation time	↑	↓	↑	
Dispersibility rating	→	↓	↑	
Mooney viscosity	↑	↑	↑	
Extrusion die swell	↓	↓ slightly	↓	

↑: increases; ↓: decreases; →: no significant change; \ goes through a maximum

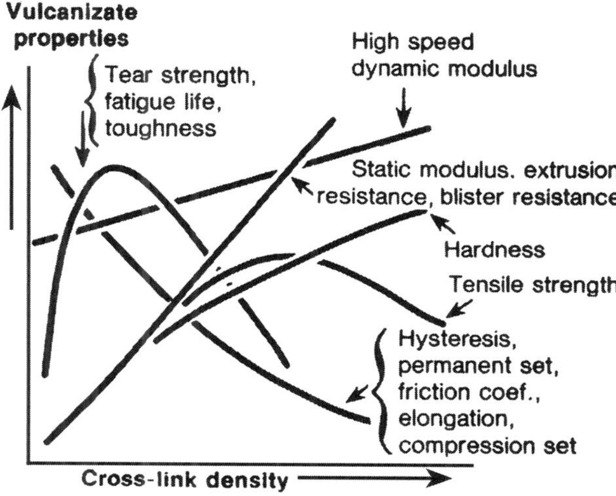

Figure 5.1: Vulcanizate properties versus cross-link density

TABLE 5.3   Effect of increasing molecular weight

	Increasing molecular weight	
Hardness	→	
Dynamic modulus	↑	Slightly
Static modulus	↑	Slightly
Tensile strength	↑	
Elongation	↑	
Compression set	↓	(Improves)
Tear strength	↑	
Fatigue life	↑	
Abrasion resistance	↑	
Impact strength	↑	
Heat build up due to hysteresis	↓	(Improves)
Extrusion resistance	↑	
Blister resistance	↑	
Electrical conductivity	→	
Processability	↓	

↑: increases; ↓: decreases; →: no significant change

fatigue life and toughness, all important requirements for oil field seals, pass through an optimum at low cross-link density and then fall off whereas sealing properties, such as hysteresis and compression set, improve with increased cross-linking.

It should be noted that hardness, modulus, extrusion resistance and resistance to blistering under rapid decompression can all be increased with an increase in cross-link density. However, the elongation is reduced, the elasticity diminished and so the sealing potential is lessened.

STRAIN OR THE STRETCHING CRYSTALLIZATION

The strain or the stretching crystallization of rubbers is another important property to be considered in sealing applications. The stretching crystallization of various rubbers is given Table 5.4.

In some oil field applications (deep wells), where temperatures approach 500°F (260°C) and pressures of 20 000 psi are reached, it would be useful to have strain crystallizing rubber seals made out of natural or neoprene rubber. At these pressures, however, the extrusion gap must be closed and the rubber seal must be protected by a series of harder and high modulus rubber back-up rings. Properly designed, this progression of material prevents extrusion, tearing and cutting or other destructive deformation of the rubber seal and distributes loads more uniformly.

FLUID TYPES IN OIL FIELD SERVICE

Sealing materials for high pressures from specialty rubbers specially compounded for oil field service are more expensive and seal design more complicated. Higher sealing pressure increases sealing force and friction. Increased friction causes higher wear rates and may require more frequent seal replacement, but frictional force and wear rates typically increase rather slowly. The fluid resistance of rubber sealing rings is directly related to the chemical nature of the fluids being handled and the chemical nature of the rubber being used. The general fluid types encountered in the oil industry are given below:

1. acids: strong or weak

2. alkali: strong or weak

3. hydrocarbon fluids

   - aliphatic – kerosene etc

   - aromatic – benzene, toluene etc

TABLE 5.4   The stretching crystallization of elastomers

Name	Chemical name	Vulcanizing agent	Stretching crystallization	Gum tensile strength
Natural rubber	CIS-1,4-polyisoprene (>99%)	Sulfur	Good	Good
Styrne butadiene rubber	Poly butadiene–co-styrene	Sulfur	Poor	Poor
Butadine rubber	Polybutadine CIS-1, 4(>97%)	Sulfur	Fair	Fair
Isoprene rubber	CIS-1, 4-polyisoprene (>97%)	Sulfur	Good	Good
EPDM	Polyethylene–co-propylene-co-diene	Sulfur	Poor	Poor
Butyl rubber	Polyisobutylene-co-isoprene	Sulfur	Good	Good
Nitrile rubber	Polybutadiene-co-acrylonitrile	Sulfur	Poor	Poor
Chloroprene rubber	Polyacroprene (mainly trans)	Mgo or Zno	Good	Good
Silicones	Polyalkylsiloxane (mainly polydimethyl-siloxane)	Peroxides	Poor	Poor
Fluorocarbon elastomers	Polyvinyledene fluoride-co-hexafluoropropene	Diamines	Poor	Poor
Polysulfide rubber	Polyalkylene sulfide	Metal oxides	Fair	Poor
Polyurethanes	Polyurethanes	Di-isocynate	Fair	Good
Hypalon	Chlorosulfonated polyethylene	Sulfur	Varies with proportions of chlorine and sulfonyl groups	Varies with proportions of chlorine and sulfonyl groups

Source: Morton, M. In: Kirk-Othmer's encyclopedia of chemical technology, 4th edn, (Kroschwitz, J.I., Howe-Grant, M., eds), Vol. 8, pp 906–907. Wiley-Interscience, New York

- chlorinated – degreasers etc

- oxygenated – ketones, ethers and esters

4. lubricants and hydraulic fluids

- low aniline point mineral oil – aliphatic

- high aniline point acid – aromatic

- water

- glycol

- silicones

- silicate esters

- diesters

- phosphate esters.

Charts and data from manufacturers and suppliers, that cover properties, ratings, and resistance to chemicals and fluids will be useful in finding out the reactivity and fluid resistance of a particular rubber.

Plasticizers, process aids, tackifiers etc. are broad subjects which can only be mentioned here in the context of seals for oil field use. Full details of the structure and activity of these ingredients, and the theories of their reaction mechanisms when incorporated in rubber compounds, cannot be treated in this book.

Plasticizers, process aids and tackifiers can offer the following advantages in seal compounds:

1. Lower cost

2. Improved low or high temperature properties

3. Easy release from molds

4. Improved tack

5. Improved extrusion profiles

6. Lower hardness

7. Lower modulus.

More information and technical data about these products can be sourced from the respective suppliers, and this information can be used to determine the particular advantages or disadvantages of each material.

PHYSICAL PROPERTY TRENDS

It is virtually impossible to achieve higher hardness without higher modulus. When tensile strength goes through a maximum, there is a decrease in elongation. Hardness increase can be achieved by increased cross-link density or by incorporation of selective fillers.

Compatible property groupings are suggested by Mastromatteo et al. [5], and reproduced in Tables 5.5–5.8. They are based on the property trends observed when filler levels and cross-link densities are changed. Table 5.5 shows that, a rubber seal functioning at low pressures will be of low hardness. The product is expected to have low modulus, low to intermediate tensile strength, high elongation etc. These are the facts to be borne in mind when designing compounds for seals and 'O' rings.

Table 5.6 shows expected properties when sealing at intermediate pressures is required. This is a very broad group.

Table 5.7 shows anticipated properties when high pressure sealing is required.

Table 5.8 shows incompatible properties, which are very rarely desired.

TABLE 5.5 Compatible properties, group 1

Low sealing pressure capability	Low to intermediate tear strength
Low hardness (40-60, A)	Good fatigue life
Low dynamic modules	Low to intermediate abrasion resistance
Low static modules	High impact strength
Low to intermediate tensile strength	Low to intermediate buildup due to hysteresis
High elongations	Low extrusion resistance
Intermediate compression set	Low blister resistance

TABLE 5.6 Compatible properties, group 2

Intermediate sealing pressure capability	Intermediate tear strength
Intermediate hardness (60-85, A)	Good fatigue life
Intermediate dynamic modules	Good abrasion resistance
Intermediate static modules	High impact strength
High tensile strengths	Low heat buildup due to hysteresis
Intermediate elongations	Intermediate extrusion resistance
Low compression set	Intermediate blister resistance

TABLE 5.7 Compatible properties, group 3

High pressure sealing capability	Poor fatigue life
High hardness (85, A→D Scale)	Good abrasion resistance
High dynamic modules	Low/intermediate impact strength
High static modules	Intermediate/high heat buildup due to hysteresis
Good tensile strength	High extrusion resistance
Low elongation	High blister resistance
Low intermediate compression set	

TABLE 5.8 Incompatible property request

A	C
1) The best tear strength and fatigue life	1) Highest hardness
2) The best compression set resistance	2) Highest elongation
B	D
1) Lowest hardness	etc.
2) Highest strength	

The information in the tables is intended to act as useful guidelines rather than written rules or standards in designing rubber compounds for sealing applications. However, the following conclusions can be drawn:

1. The rubber chemist should optimize every variable, including cost, in order to achieve a dependable, successful rubber seal.

2. For every property achieved, some other will be lost. Compounding plays a secondary role in setting desired properties.

3. High molecular weight rubbers are the best choice for strength at high temperatures.

4. Cross-link density and filler choice (based on type, quantity, and choice of plasticizers) are key factors for property optimization.

5. Heat and fluid resistance are closely linked to the chemical composition and structure of a rubber.

6. Use of a rubber selection guide should be the first step, since most of the critical properties of a rubber are intrinsic.

REFERENCES

1 The World Intellectual Property Organization (WIPO) http://www.wipo.int/about-wipo/en/what/
2 Schlumberger website: ttp://www.slb.com/content/about/index.asp?entry=about&
3 http://www.glossary.oilfield.slb.com/Display.cfm?Term=stimulation%20fluid)
4 Briscoe, B.J., Zakaria, S. et al Role of interfacial quality on gas induced damage of elastomer composites. Presented at Explosive Decompression Seminar, June 4, 1990 Red bank NJ, USA. Dan Hertz, Seals Eastern Inc; Energy Group Educational Symposium, September 24–25, 1991
5 Mastromatteo, R., Morrisey, E., Mastromatteo, M.E., Day, H.W. Matching material properties to application requirement. Rubber World 187 (5), 1983

CHAPTER 6

RUBBERS, CHEMICALS AND COMPOUNDING FOR 'O' RINGS AND SEALS

DESIGN OF COMPOUNDS FOR 'O' RINGS

A designer of a rubber compound needs to note the following three requirements:

1. End-use and service requirements

2. Ease off processing

3. Volume cost and availability of raw materials.

Various types of rubbers and compounding chemicals can be specified in a formulation to meet specific requirements.

Natural rubber will almost certainly be used in cases where the compound does not have to withstand high temperature, direct sunlight or ozone, and where it does come into contact with oils, solvents, fluids or chemicals. Synthetic rubbers, which have better resistance than natural rubber to these factors, are used where they will be encountered routinely. Typical applications of some synthetic rubbers are given below:

- styrene butadiene rubber: general purpose applications, replacing natural rubber.

- butyl rubber: where low permeability and good chemical resistance are required.

- butadiene rubber: used for good abrasion and fatigue resistance and in blends with natural and styrene butadiene rubbers. Tear resistance is poor.

- nitrile rubber: where good oil and solvent resistance is needed.

- chloroprene rubber: has good ozone and flame resistance.

Rubber Seals for Fluid and Hydraulic Systems 2010; ISBN: 9780815520757

RUBBER BLENDS

Rubbers are often blended with one another, or copolymerized with another in the polymerization stage. The blends are expected to have better properties and price than, the individual components hence giving technical and economic advantage. For example, the addition of small amounts of a selected synthetic rubber to another may improve such properties as oil and ozone resistance, or improve processing behavior.

Blends in emulsion and in solution synthetic rubbers polymerized often give processing difficulties due to their differences in Mooney viscosities and curing characteristics. In this case the final properties of the blend will be inferior to levels attainable with the individual rubbers. Properties adversely affected by non-compatible rubber blends include tensile strength, permanent set, low temperature behavior and covulcanizability. Reactivity differences in the blended phases, and/or diffusion of vulcanizing agents (which are generally polar) from the less polar to the more polar phase, may mean that consistency of blending may be difficult to achieve under factory conditions. Inadequate covulcanization then leads to unfavorable mechanical properties, such as low tensile strength and poor dynamic behavior. This is also true when blending different grades of the same synthetic rubbers which have different Mooney viscosities and curing characteristics.

Masterbatches

A masterbatch is a mixture of rubber with one or more compounding ingredients in higher concentrations than would normally be present in a complete compound. Masterbatching is used for accuracy and efficiency in dispersion during mixing, in order to avoid adding accelerators, antioxidants, pigments, sulfur and peptizing agents in small quantities. Sulfur masterbatches and peptizer masterbatches are very common in compounding practice.

Choice of rubber

Natural rubber (NR)

Natural pure gum stocks can be compounded for tensile strengths which are quite high compared to those of non-crystallizing rubbers. This is associated with the tendency of natural rubbers to form crystallites when strained. Tensile strengths can be further enhanced (within limits) by an increase in cross-link density, or at constant cross-link densities by the addition of reinforcing carbon blacks. Reduced compression set can also be achieved by raising the cross-link density although this gives a marginal reduction in tensile strength, fatigue and aging properties. In natural rubber compounds, silica is often added to provide high tear strength without a prohibitive increase in heat build up. Semi-efficient vulcanization systems can give improved resistance to reversion and compression set at elevated temperatures. Air permeability of natural rubber compounds can be improved by the use of NR/CR (natural rubber/chloroprene rubber) blends.

Styrene Butadiene Rubber (SBR)

Styrene butadiene rubber is generally marketed at lower viscosity grades than NR and this permits its use in rubber compounding without premastication. Mechanical or chemical peptizing (or dispersing as a colloid, or suspension) is not required in SBR rubber. While most properties of SBR are comparable with NR, in some respects, such as heat build up, tack and gum tensile strength, SBR is inferior but addition of resins and reinforcing fillers improves these properties acceptably.

Owing to its irregular molecular structure, SBR does not crystallize, so reinforcing filler is required to achieve good physical properties. The best of these is carbon black; other fillers give very inferior properties in SBR to those achievable in a corresponding natural rubber compound. Resistance to abrasion, and degradation under heat is better for SBR, but: as it is less chemically reactive it is also slower to cure, requiring more accelerators. Scorch problems are less likely to occur with SBR than with NR especially with carbon filled compounds. It is compatible with NR, butadiene rubbers, ethylene–propylene terpolymer (EPDM), nitrile and chloroprene rubbers. The effect of the sulfur accelerator ratio on the cross-link distribution in SBR is similar to NR. In general, a lower sulfur level and a higher accelerator level are normally employed for SBR cure systems.

Emulsion polymerized SBRs are used alone, as well when as blended with butadiene or natural rubber. In addition with other diene rubbers, higher levels of filler (75 phr of carbon black and plasticizer 20 phr, or more than the quantity contained in oil-extended SBR) may be used.

Cold polymerized SBR is used in light duty products, such as hoses, or molded goods. Its low resilience in black-filled compounds restricts its use in products where high heat build up is likely, such as static seals. Antidegradants can be added to SBR to provide protection during service. Building tack can be improved by the addition of tackifying resins, or by blending with natural rubber. The oil resistance of SBR can be improved by blending nitrile rubbers into the compounds. Petroleum and coal-tar type plasticizers are used to give easier processing, and reduced costs. Solution polymerized SBRs are generally more expensive than the emulsion types, and so are used in specialty applications, where higher glass transition temperature (Tg) and grip while in contact with fluids are required. This is particularly useful in high performance rotary seals.

Polybutadiene rubber (BR)

A blend of 20–30 percent high cis-4- polybutadiene and 70–80 percent of NR is used in tires because of its high abrasion resistance and low heat build up. Although this trait is very desirable in oil-resistant 'O' rings, this rubber cannot be used here, because it is adversely affected by fuels and hydraulic fluids. As it has limited compatibility with polar rubbers, such as acrylonitrile, it can only be blended with them at levels of up to 25%. Blends of poly-butadiene rubber are used partly due to its poor processing behavior as a homopolymer; it does not respond well to mastication, it does not form a smooth band on a mixing mill, and it produces extrusions of rough appearance. It is therefore advisable to mix all compounds containing high proportions of BR in internal mixers.

Curing of BR may be achieved by using conventional sulfur systems activated with metal oxides and fatty acids like stearic or oleic acids. In comparison with NR, BR requires less sulfur but marginally higher accelerator levels, to achieve similar levels of cross-link density.

Acrylonitrile butadiene rubber (NBR)

Nitrile rubber has been the work-horse elastomer in oil and gas production for many years, but as oil wells began to produce increasing amounts of hydrogen sulfide (through stimulation or reservoir conditions coupled with higher production temperatures), these seals began to fail. Over the past years, nitrile rubber has become the elastomer of choice for the oil and gas industry, as it is durable at high temperatures, and resistant to chemical attack. However, as wells become deeper, production temperatures inevitably increase, cuusing the rubber to harden, and the seal to fail.

Acrylonitrile butadiene rubber rubbers do not crystallize under strain and, without reinforcing fillers, have poor tensile strength and low tear initiation resistance. Therefore, it is usual to have reinforcing fillers like carbon black, phenolic resins, or polyvinyl chloride (PVC) to improve tensile strength. Very hard compounds can be prepared by using phenol formaldehyde (PF) resins with NBR/PVC blends.

Although reinforced NBR vulcanizates have lower tear resistance than NR compounds, the fall off at higher rates of deformation and higher temperatures is smaller it also shows less abrasion loss than NR.

The incorporation of 30 phr (parts per hundred) PVC into NBR compounds results in improved weather resistance and high resistance to swelling. The presence of PVC, however, produces stiffer compounds, with lower resilience and inferior low temperature properties. To achieve optimum results, complete gelling of PVC with NBR at 160°C or higher temperatures, in the presence of an effective PVC stabilizer, such as a metallic soap, is necessary.

Acrylonitrile butadiene rubber is susceptible to ozone attack, and therefore suitable anti-ozonants and/or waxes need to be added. Acrylonitrile butadiene rubber /EPDM blends (at 70/30 or 80/20 ratio) have a better ozone resistance than NBR alone.

The heat resistance of NBR can further be enhanced by the use of low sulfur or sulfur donor curing systems. Plasticizers of low volatility should be chosen to produce heat stable compounds.

Because of its strongly polar nature, NBR is compatible only with polar plasticizers, such as dioctyl phthalate, dioctyle sebacate, esters of carboxylic acid, and phosphoric acids.

Acrylonitrile butadiene rubber has low permeability to gases, depending on its acrylonitrile content. The high polarity of NBR also influences its compatibility with certain compounding ingredients, and non-polar rubbers such as SBR and BR. Good adhesion between NBR and textile fabrics can be achieved with NBR lattices or 50:50 blends of NBR and vinyl pyridine lattices in resorcinol formaldehyde dips. Generally, non-staining or slightly staining stabilizers are used in the manufacture of NBR.

Acrylonitrile butadiene rubber is a semi-conducting rubber which can be made fully conductive by the addition of FEF (fast extrusion furnace) blacks and acetylene carbon blacks.

Antistatic compunds can be prepared by the addition of plasticizers.

The low temperature flexibility of NBR can be improved by the addition of dioctyl adipate (DOA) or dioctyl sebacate (DOS) type plasticizers.

Isobutylene–Isoprene Rubbers: (IIR) Butyl rubber

Butyl rubbers, unlike many other rubbers, do not break down easily during the normal mixing process in the mill. The compound viscosity can be controlled by selection of the proper grade of butyl rubber, the type of carbon black and the type and loading level of oil. Relatively high loading levels of paraffinic or naphthenic oils are employed for typical butyl applications, such as inflatable seals, tubes or sealing bladders and solid body items.

Butyl gums have high tensile strengths, and this is reduced by the addition of filler 50–100 phr of carbon blacks, either GPF (general purpose furnace) or FEF or 100–150 pfr mineral fillers, such as hard clay, talc or silica, are often added if a lower tensile strength is required. The preferred plasticizers are highly saturated materials of relatively low polarity, e.g. hydrocarbon oils, waxes etc. Heat treatment procedures, in which butyl rubber and carbon blacks are subjected to high temperatures (160–200°C) for at least 2–3 minutes generally give improved vulcanizate properties in the presence of dinitroso benzene. Butyl vulcanizates are highly damping and this is more pronounced at lower temperatures.

Maximum levels of ozone resistance can be achieved when the lowest unsaturation grades of butyl rubber are vulcanized to the highest possible state of cure. Plasticizers reduce ozone resistance, but excessively high loadings of carbon blacks or light colored filler loadings should be avoided. Protecting agents are not usually necessary, but wax or blends with EPDM rubbers can be employed to advantage.

Strong acids and bases, and strongly oxidizing or reducing materials do not attack butyl rubber, however, concentrated nitric or sulfuric acids can cause degradation. In compounding for minimum swell in acids, high filler loading, high reinforcement and maximum state of cure are very important. Chemically sensitive fillers, such as calcium carbonates, should not be used in compounds coming in contact with mineral acids.

Carbon loaded butyl vulcanizates have outstanding weathering properties. To protect non-black butyl products from sunlight, it is desirable to use adequate loadings of high opacity pigments, such as titanium dioxide at 15 phr, or zinc oxide at 45 phr.

The inertness of polyisobutylene chains in butyl rubber gives outstanding resistance to heat and oxidation. However, low sulfur, sulfurless, or sulfur donor systems having more thermally stable cross-links, show better aging effects than conventionally sulfur-cured products. Resin cures are extremely stable towards heat and oxidation, and these vulcanizates are preferred for high temperature use. Heat-treated butyl rubber compounds result in good carbon black dispersions, and in better dimensional stability during extrusion and calendering.

Halobutyl rubbers: BIIR and CIIR (bromobutyl and chlorobutyl rubbers)

For halobutyl rubbers, zinc oxide is used as the vulcanizing agent with low levels of sulfur stearic acid and sulfenamide accelerators. Bromobutyl rubber shows higher cure reactivity than chlorobutyl rubber. Both CIIR and BIIR will cure with zinc oxide, but only BIIR will cure with sulfur alone, no zinc oxide or accelerator being necessary. Bromobutyl rubber can be cured with 0.5 phr of sulfur and 1.3 phr of dibenzothiazyl disulfide accelerator, 3 phr of zinc oxide and 1 phr of stearic acid. Levels of sulfur as low as 0.5 phr will give a rapid and reasonable degree of cure. Zinc diethyl dithiocarbomate (ZDC) accelerator can be used in small quantities (0.25–0.75 phr) with zinc oxide for heat resistant products, and to improve compression set.

Peroxide vulcanization is also suitable for BIIR. While IIR degrades rapidly when heated in the presence of organic peroxides, BIIR can be cured with peroxides in combination with a coagent. Vulcanizates with unusually low compression set, heat resistance and excellent ozone resistance can be produced with peroxide cures.

Magnesium oxide and calcium stearate, when used along with the above accelerators in halo butyl rubbers, act as retarders. They also have an effect on cure time and state of cure, as well as increasing scorch time.

Pre-cross-linked butyl rubber

This product is a terpolymer of isobutylene isoprene and divinylbenzene, the latter giving a high degree of pre-cross-linking. The pre-cross-linking increases green strength, provides dimensional stability and reduces cold flow. This rubber is used as sealant.

EPDM Rubbers

These are amorphous terpolymers. Like many other non-crystallizing rubbers, mechanical properties of unfilled EPDM rubbers are rather poor and, consequently, reinforcing fillers are incorporated in the compounds. The extraordinary mechanical properties that are obtainable with high reinforcing fillers are not required in specialty applications, such as seals and gaskets, when of EPDM rubbers are used. Use of semi-reinforcing carbon blacks in EPDM compounds is therefore common. Its low density makes it possible to use high extender oil and filler loadings for economy, without adversely affecting performance characteristics. Oil extended EPDMs are used for compounds of low hardness. For harder compounds, blends with NR, SBR, high styrene resins or phenolic resins are used.

Generally, an EPDM cure system will contain a thiazole (mercaptobenzothiazole or MBT, dibenzothiazyl disulfide or MBTS etc.) in combination with a thiuram and/or a dithiocarbomate. For higher heat resistance, the sulfur donor accelerators may replace part or all of the elemental sulfur in the cure package, as exemplified below:

MBTS – 3 phr

TMT – 0.8 phr

ZDC – 1 phr

Sulfur – 0.5–0.7 phr.

The main chains of the EPDM molecules have no double bonds, and thus EPDM does not suffer deterioration due to molecular scission, even after extended exposure to sunlight or high concentrations of ozone. This makes them resistant to weathering. The compounds based on EPDM rubbers are generally suitable for severe outdoor conditions or high ozone concentrations, without the need for antiozonants, waxes etc. General purpose diene rubbers, such as NR and SBR, which are not as good in this respect are therefore blended with EPDM. Although they do not possess the excellent low air permeability of the butyl rubbers, they are used at fairly low levels in inflatable seals to improve low temperature properties, and to give good aging and crack-resistant characteristics.

There are a number of modifications which can be made to EPDM rubbers. The different commercial grades currently available differ in how they are polymerized, that is by solution or suspension polymerization. They also differ in Mooney viscosity, molecular weight distribution, ethylene–propylene ratio, and type and amount of third monomer (DCP or ENB). Suitable grades are chosen by considering the process to be used, application requirements and costs. Ethylene–propylene copolymer (EPM) does not have double bonds in the polymer. Hence, peroxides are used for curing, and the resulting vulcanizates excel in heat resistance. EPDM can be cross-linked with sulfur or peroxide systems as it has an unsaturated diene group on the termonomer side chain. Peroxide cure systems may be necessary if the application requires resistance to temperatures of 150–170°C and a very low compression set. Blending with NR and SBR rubbers is facilitated by the use of high diene EPDM (i.e. having a high level of unsaturation). Commercial grades of EPDM contain a maximum of 15 double bonds per 1000 carbon atoms, which contrasts with SBR or NR which have 150–250 double bonds per 1000 carbon atoms in the main chain. Thus, resistance to ozone, heat, UV radiation, and humidity is excellent. EPDM can be considered as a non-staining, non- migrating antioxidant and can be used as a 15–30% blend with natural rubbers, or SBR. Adhesion of EPDM to textiles is fairly poor, but can be achieved by using suitable bonding agents, or by incorporating SBR or vinyl pyridene lattices.

EPDM is an excellent radiation-resistant rubber, and it has good low temperature flexibility, staying flexible to temperatures of $-50°C$. This rubber is resistant to polar fluids, alcohol, glycol, ketone and phosphate esters, acids, alkalis, salts and fats. Compounds from EPDM can be used where chemical resistance is required. They do not possess resistance to non-polar hydrocarbon oils and solvents.

Fluorocarbon rubbers

The most important use of fluorocarbon rubbers is for the production of seals, such as gaskets and 'O' rings, for sealing fluids. They have high resilience and general impermeability to many fluids. This is important not only to avoid the loss or contamination of valuable materials, but also to protect personal safety, health and the environment.

A good seal should have many properties. Firstly, it must not degrade, or lose its elasticity in contact with the fluids being handled, even at extremes of temperature and pressure. It must not become permeable to these fluids, react with them or contaminate them with impurities. The seal must not swell appreciably when in contact fluids, even at elevated temperatures. For example, if an 'O' ring in contact with a moving surface swells appreciably, it may be extruded into a close-clearance area where it will be torn by friction. When the pressure or temperature is then reduced, the 'O' ring may shrink from its extruded position in a way that causes fluid leakage.

Fluororubbers have been utilized extensively for sealing applications because of their outstanding chemical inertness, solvent resistance, and their resistance to high temperatures. In such applications, their high cost is readily justified by their longer life and superior performance.

Existing fluoroelastomers are not suitable by themselves for sealing certain organic fluorochemical fluids. Since they are chemically similar, the fluids tend to dissolve in the fluoroelastomers causing swelling and physical deterioration. When such fluorochemicals are manufactured, many of the process streams also contain chlorocarbons highly corrosive hydrofluoric acid (HF), or hydrochloric acid (HC1), making the situation worse. In particular, 2,2-dichloro-l,l,l-trifluoroethane (HCFC-123) is one of the most difficult products for which to find a satisfactory seal. HCFC-123 is a synthetic, non-combustible, volatile liquid that is used as a refrigerant in commercial and industrial air-conditioning installations, in gaseous fire extinguishers, as a foam-blowing agent and in metal and electronics cleaning. There is a definite need in the fluorochemical industry for an elastomer that is suitable for sealing fluorocarbons, chlorocarbons, hydrogen fluoride, hydrogen chloride and mixtures thereof.

Fluorocarbon rubbers, or fluoroelastomers are a versatile class of materials. They have unique combinations of physical properties that are capable of providing premium performance in sealing even aggressive agents such as HCFC-123, and handling and controlling fluids and gases under extreme conditions. Fluorocarbon rubbers are hydrocarbon polymers in which significant amounts (50–70%) of the hydrogen has been replaced by fluorine gives them their exceptional resistance to attack by hydraulic fluids, fuels, oils and corrosive chemicals. The high fluorine content is also responsible for their excellent retention of physical properties after exposure to air at elevated temperatures, and over extended periods of time. They are also mechanically tough, and have excellent resistance to hot oils, synthetic lubricants, gasolines, fuels and many organic solvents. In addition, they can be selectively compounded for resistance to hot mineral acids, steam, hot water and a number of organic acids.

A typical 'O' ring formulation to produce a fluororubber of Mooney viscosity of 40 at ML1 + 10 at 250°F (121°C) can be:

Fluororubber – 100 phr

Magnesium oxide – 3 phr

Calcium hydroxide – 6 phr

Medium thermal black – 30 phr

It is likely to have the following physical properties:

- Tensile strength – 1780 psi

- Elongation at break – 195%

- Shore hardness – 75°A

- Compression set

- after 70 h at 392°F (200°C) – 14%

- cure time – 5 minutes at 350°F (175°C)

- post cure – 24 h at 500°F (260°C).

Fluororubbers can be used over a temperature range of −50°F (−45.5°C) to + 600°F (315°C), and can be rated for continuous service at 450°F (230°C), without any significant loss of either molecular structure or mechanical properties. The best property of fluororubber is its compression set resistance, highly desirable in fluid sealing applications. For 70 h at 400°F (200°C) this reach levels of 10–20% the lowest of any rubber.

While most fluororubbers will burn in direct flame, they self-extinguish when the flame is removed. They are highly-weather and ozone resistant, and show no change in physical properties when exposed to the atmosphere for long periods.

Compounding advance in fluororubbers, by various manufacturers, have widened their versatility by increasing hardness from 50° shore A to 100° shore A, and improving tear resistance, elongation and compression set. Fluororubbers were commercialized in the 1960s and, since then, significant improvements have been made in processibility, aging characteristics and physical properties.

One of the largest growth areas for fluororubber is in the automobile and aviation industries. For example, front engine, and front and rear crank shaft seals are molded from fluoroelastomers, and although unleaded gasoline will make most rubbers swell, it does not affect fluororubbers. Such characteristics make fluororubbers the best and, at times, the only choice for critical seal applications. Autotransmission cable seals are made extensively from this material, as are many seals for diesel engines.

Although expensive, fluororubbers have excellent processing characteristics. It is a low viscosity polymer, so flows better and eliminates air pockets that are trapped during forming and molding operations. It does not make the molds dirty and can be formed into unusual shapes, such as seals with extended lips. It also bonds well with other substrates. Because of these special characteristics, seals or 'O' rings made with fluororubbers vary very little, easing quality control, and as the scrap rate is low, manufacturing costs are comparatively low. A thin ring of polytetrafluoroethylene (PTFE) is used where the seal touches the shaft to minimize friction and prolong life. Typical automobile applications of fluororubber components and seals are shown in Figure 6.1.

Ethylene acrylic rubber

Ethylene acrylic rubber is manufactured by M/s Dupont USA under the trade name of Vamac, and is about half ethylene and half methylacrylate. A small amount of cure site monomer in the molecule provides the ability to cross-link chemically. This rubber is the combination of two major chemicals which give its unique balance of properties. For instance, the backbone structure of the polymer molecule is saturated, and so it is inherently resistant to ozone attack. The acrylic segment provides oil resistance, and the ethylene segment yields low temperature performance. The added feature of this rubber is that there is no halogen present to become corroded. There is slightly more tendency to swell than a homopolymer, such as polyacrylate or acrylonitrile rubber, but it is approximately equal to silicone, chloroprene and Hypolan (chlorosulfonated polyethylene) rubbers.

Fluid exposure data for some automotive fluids are shown in Table 6.1 for this rubber. Such data for conventional synthetic rubbers is not provided since it can be sourced in manufacturers' trade brochures.

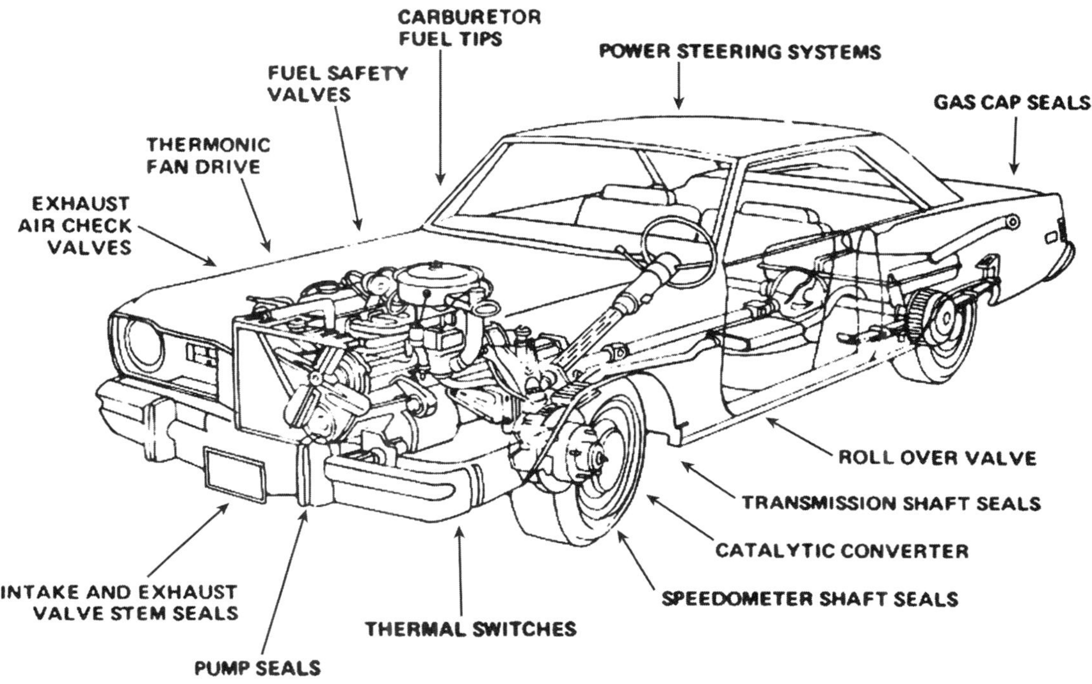

Figure 6.1: Typical automobile details of fluoroelastomer parts. Source: *Rubber World* 180(1) 1979

TABLE 6.1 Fluid resistance

Fluid	Temperature (°C) (%)	H	Volume increase
ASTM oil no1	150	70	3
ASTM oil no 3	150	70	50
Axle lubricant	150	168	15
ATF (aviation turbine fuel)	150	168	16
Freon	24	168	30
ASTM reference fuel C	24	168	150
Brake fluid	121	168	165

Silicone rubber

The few widely used silicone rubbers are polydimethylsiloxanes, polydiphenylsiloxanes and polymethyl–phenylsiloxanes, collectively called silicones. With a repeating unit of silicon–oxygen, the siloxane chemical backbone structure possesses excellent thermal stability and flexibility, superior to most other rubbers. Polydimethylsiloxanes provide a very low glass transition temperature (Tg), but the rubber can be used at temperatures up to 200°C.

Low values of tensile strength are common with unreinforced silicone rubbers, so reinforcement is necessary in silicone compounding. Unlike most organic rubbers, they can be either very difficult or impossible to cross-link with sulfur cure systems. The main chemical routes for vulcanization of silicone rubbers are:

1.   Elevated temperature vulcanization

2.   Room temperature vulcanizing mechanisms (RTV).

In elevated temperature curing systems, organic peroxides are mostly used. However, they are inhibited by most carbon blacks and so non-black reinforcing fillers, such as precipitated silicas, titanium dioxide and zinc oxide are used. Room temperature vulcanization is normally adopted for low consistency silicone rubbers. The material can then be easily extruded and then vulcanized. Since silicone rubbers are used in various application areas, they are available in the market in various grades of specific properties, such as high strength and low compression set. The rubbers suitable for low compression set applications, such as seals or 'O' rings, do not require elevated temperature post cures following initial vulcanization. Silicone rubber is very useful for high temperature services, has excellent weathering properties, thermal stability, ozone and oxidation resistance, good electrical properties, and extreme low temperature flexibility. They are, however more permeable to gases than natural rubber.

Chloroprene rubber (CR)

These rubbers, known commercially as neoprenes, are available in different grades, classified by Mooney viscosity, crystallization tendency, degree of pre-cross-linking and modifier type used in the polymerization process. Due to its structural uniformity, the CR vulcanizates from pure gum compound or compounds have higher tensile strength than those from similar NR-based vulcanizate, especially if inert fillers are used alone. A number of variables can be changed when processing CR vulcanizates, which can alter their chemical composition. This is useful when aiming for a given tensile strength in the product. Chloroprene rubber compounds can be formulated by following procedures for NR and SBR fillers and plasticizers. This rubber is vulcanized with metal oxides, either with or without sulfur; usually a combination of zinc and magnesium oxides, with white lead oxide (litharge or red lead). Lead oxide gives good results where low water absorption is an important requirement. The sulfur modified grades require no additional accelerator for vulcanization. In the case of mercaptan modified grades, ethylene thiourea increases the rate of vulcanization and the modulus, but reduces the storage stability of the compound. The choice of antioxidants is also peculiar to CR. Diphenylamine derivatives or paraphenyledene diamine antidegradants, with ozone protective waxes, are used. Quinoline type antioxidants should be avoided.

Antioxidants or antidegradants are generally only used in rubber compounding for specialty applications such as oil seals, or where chemical and wear resistance is required. The benefit of such protection is rarely seen in practice, however, as it is likely to fail due to the effect of radiation or chemical attack before oxidation progresses very far.

High tear strength can often be obtained by using reinforcing silica fillers. Other fillers can be used to enchance chemical resistance, such as antimony trioxide, zinc borate or aluminum hydroxide. Synthetic plasticizers, such as fire retardant phosphoric acid esters, chlorinated waxes etc, silica fillers, talc or china clay enhanced resistance to hot water, and fine particle talc enchances chemical and acid resistance. Ester plasticizers reduce the glass transition temperature and improve the low temperature flexibility. The mercaptan grades are more heat resistant than sulfur modified grades. To avoid CR compounds sticking on mill rolls, the addition of stearic acid, paraffin wax, low molecular weight polyethylene, or polybutadienes as processing aids at levels of 0.5–1.0 phr is often necessary. The temperature of compounds without accelerators should not be allowed to exceed 130°C during processing, as CR shows a tendency to cyclize at higher temperatures. The effect of heat is cumulative and results in progressive reduction of processing safety. It is therefore suggested that mills at low friction ratios (1:1.1 or 1:1.2) should be employed, in order to minimize heat build up.

Chloroprene rubber has a combination of several technically essential properties, such as good mechanical properties, heat resistance, ozone and weather resistance, flame resistance and adequate electrical properties, which are not obtained with other synthetic rubbers. Thus, it has a broad spectrum of usage.

Reclaimed rubber

Apart from its role as a cheapening extender, reclaimed rubber, in many cases, aids processing considerably by reducing the nerve of the rubber. It is also a useful ingredient in stocks required for bonding to metal as it confers improved adhesion. Batches containing reclaim are mixed more easily and rapidly and, where high levels of physical properties are not necessary, it can be used in large quantities. The rubber chemist balances performance, cost and ease of processing before making use of reclaim as it can adversely affect physical properties such as resilience, fatigue and abrasion resistance if used in high proportions. The loss of required physical properties where reclaimed rubber is used means that it is not used in the manufacture of 'O' rings and seals. It is, however, used in many automotive and industrial molded products and rubber sheets.

CHAPTER 7

RUBBER EXPANSION JOINTS

Piping is a major component of all chemical processing plants. The piping engineer is concerned with stress levels and movement at significant points within the system, and the magnitude and direction of forces exerted on equipment connected at terminals or starting points. When stresses exceed allowable limits, fluid leaks and a drop in system pressure may occur, hence modification of the piping arrangement is needed. The simplest method of reducing stresses is to provide additional pipe fittings in the system, in the form of loops and bends. Stresses lead to movements of the piping system, so movement absorbing devices are incorporated in the layout. They are usually called expansion joints, and they are either made of rubber, with one or two convolutions, or metal, with multiple convolutions. Their purpose is to provide the flexibility to absorb movements due to thermal changes, and to absorb dynamic movements of machinery, adjoining buildings etc.

A recent report [1] studied by the author suggests that installation of rubber expansion joints in pipes will prevent breakage and leaking, by absorbing swings and vibrations:

1. On 20 April 2005, British Nuclear Group Sellafield Limited (BNGSL) discovered a leak from a pipe that supplied highly radioactive liquor to an accountancy tank in a part of the Thermal Oxide Reprocessing Plant (THORP) at Sellafield, known as the 'feed clarification cell'. The incident was categorized by BNGSL as '3' on the International Nuclear Event Scale.

2. In total, approximately 83 000 liters of dissolver product liquor, containing approximately 22 000 kilograms of nuclear fuel (mostly uranium incorporating around 160 kilograms of plutonium), had leaked onto the floor of the cell. That leak had begun prior to 28 August 2004 and had remained undiscovered until 20 April 2005. It is likely that the leak was relatively small until January 2005.

3. Video evidence indicated that the leak came from a pipe, identified as nozzle N5, which had completely severed at a point just above where it enters accountancy tank B (HEAT B – Head End accountancy tank V2217B). The most likely cause was fatigue failure from the swinging or swaying motion of the suspended tank, which occurred during agitation of the tank contents as part of normal operation. The motion occurred because of design inconsistencies in the later stages of the design process and during construction, together with a modification to the operational mode of the vessel around 1997, which inadequately considered the impact on pipework. The change process overlooked the effects of the tanks swinging on their suspension rods during agitation of liquor, leading to pipework fatigue.

Rubber Seals for Fluid and Hydraulic Systems 2010; ISBN: 9780815520757

In the reported nuclear reactor, the pipe line with the nozzle has no expansion joint; if one had been installed in the piping system, at the nozzle point, it would have absorbed the motion of the suspended tank.

Rubber expansion joints offer engineering advantages over metallic joints. Rubber expansion joints, which consist of flanged ends and a flexible section, can absorb within its free length more lateral movements, than any other similarly sized joint. Expansion bellows are joints that have several sections. The flexible section of the rubber expansion joint/bellows is often a single fold which, because of the inherent flexibility of the rubber, can deal with large lateral movements with low force, a phenomenon which would require multiple folds in a similar metal component.

The flexible section of a rubber expansion joint often has a single fold or arch (Figure 7.1). A double arch expansion joint is shown in Figure 7.2.

The major difference between rubber joints and metal bellows is in the way they absorb pressure loads. Circumferential loads (hoop stress) due to pressure are carried by the folds themselves in a metal bellows. In a rubber expansion joint, the convolution is incapable of resisting pressure by itself, but is supported by the adjacent tube with its internal fabric or fabric reinforcement or by the adjacent flanges. Expansion joints have integrally molded flanges, which are drilled to match standard pipe flanges. All expansion joints require metallic retainer rings behind the flanges as back up.

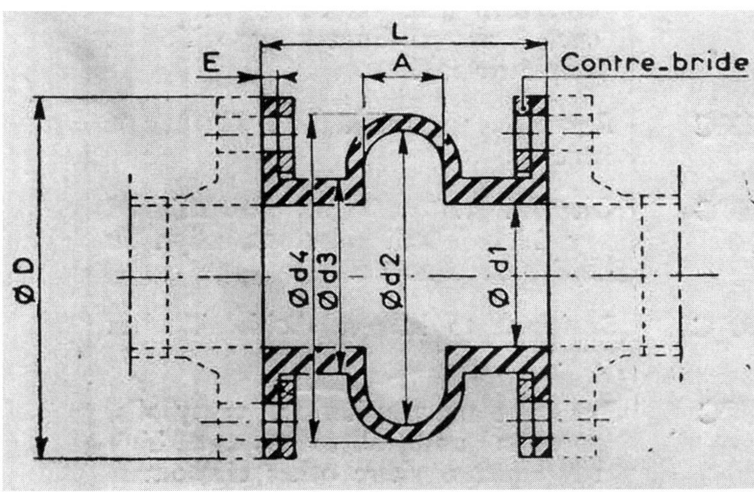

Figure 7.1: Single arch expansion joint

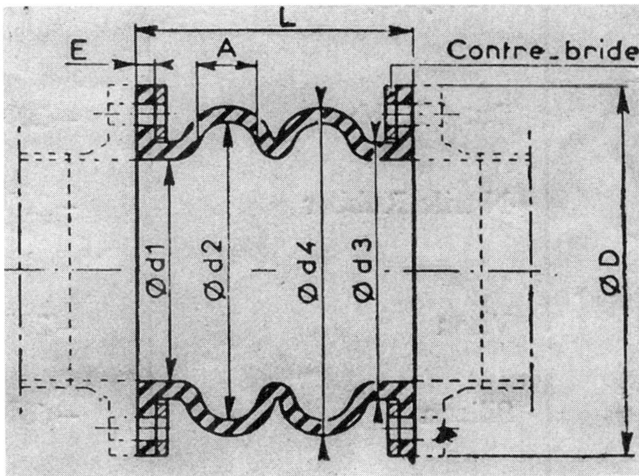

Figure 7.2: Double arch and double arch expansion joint

APPLICATIONS OF RUBBER EXPANSION JOINTS

The unique advantages of rubber expansion joints make them used in industrial settings where fluid and gas handling is a primary part of the process, e.g. waste management, or air pollution control. Wherever movements or vibrations from thermal expansion or mechanical equipment expansion occur, joints provide low cost and efficient solutions to leakage problems.

Chemical process industry

For piping systems conveying hydrocarbons, emulsions, saline solutions, air, gas, steam, liquids, hydraulic fluids and fuels, suitably designed rubber expansion joints are usually installed within the piping system to accommodate forces due to movement and thermal expansion, and so prevent leakages.

Food and beverages industry

Expansion joints are used in systems that handle drinking water and other liquids such as beer, milk, wine, vegetable oils etc.

Piping in heating and air conditioning systems

Rubber expansion joints are successfully used in several industrial plants, nuclear and military installations to control leaks in hot and cold water distribution systems, chilled water and condensed water piping, and suction and discharge sides of pumps adjacent to compressors.

Hydrocarbon process industry

Cooling systems, petroleum and gas exploration industries all use expansion joints.

The types of systems in various processing industries are widely different hence selection of suitable rubbers is made on the basis of their chemical and mechanical properties.

THE BENEFITS OF MULTIPLE BELLOWS IN INDUSTRIAL PLANTS

Bellows can vibrate both from internal fluid flow, and externally imposed mechanical vibrations. At high flow, velocities and flow induced resonance produces bellows fatigue. Multiple bellows are less susceptible to vibration failures because of the damping effect of the interplay friction. The benefits of multiple rubber expansion joints are given below:

- increase in flexibility and reducing deflection forces

- ability to cope with higher pressures and lower thrust forces

- lower spring rates and higher elasticity

- minimal installation length due to high elasticity

- fail safe design

- high corrosion resistance can be attained economically.

THE FEATURES OF A RUBBER EXPANSION JOINT

The most important feature of expansion joints is the multiple structure of the bellows elements. The bellows are manufactured as a multilayered tube of plies, ranging in number from 2 to 20, depending upon diameter and specified operating pressure.

Each of the thin layers acts as a nearly neutral fiber, which offers significantly lower resistance to movements in the piping system than a single wall could. Significantly less force is required for actuating multiple bellows.

Thus, expansion bellows can accept significantly more movement with the same dimensions. The stresses induced on the individual layers of bellows are a fraction of the stresses induced in the single wall of an expansion joint of equal thickness. This maximizes the service life of the bellows.

Advantages of rubber expansion joints

Safety

A catastrophic, sudden failure is virtually impossible when using rubber expansion joints, since the multiple inner core prevents instantaneous bursting. If a crack does occur due to stress or fatigue, the multiple plies will contain pressure, and hold the joint in place. In doing this, the plies act as a labyrinth seal, which is a mechanical seal that fits around an axle and prevents leaks. A small bubble or blister occurs in the outer ply alone, and the joint will still perform satisfactorily. Replacement can then be carried out when convenient.

Longer service life

Using multiple plies of relatively thin gauge material means that lower residual stresses are introduced when the bellows element is formed. Also, the thinner multiple plies are subjected to lower operating stresses when compared to a single heavy fabric. Bellows therefore have lower built-in and operating stresses, which ultimately results in longer life.

Compact design

Since multiple bellows have more flexibility than conventional ones, they require shorter pipe lengths to accommodate a given movement. This makes the assembly of the piping system more compact and economical.

Lower thrust forces

The multiple core design results in lower spring rates for a given pressure capacity, and the effective cross-sectional area of multiple bellows is smaller than a comparable conventional metallic bellows. These two reductions mean lower forces and moments on the anchors, equipment and guides supporting the piping system, so they can be dimensioned significantly smaller, and more economically. This increases the life of the connected equipment and raises the reliability of the overall piping system, so lowering costs.

Cost reduction

Because of higher flexibility and movement capacity, a small number of expansion joints is sufficient to absorb large displacements.

Less downtime

Even if the inner plies become damaged, the multiple expansion bellows do not lose their capacity to perform essential functions. Since the outer plies are intact, the joint can still resist pressure, and its flexibitity is retained. This gives enough time to replace the joint during routine maintenance.

Expansion and compression strains

When expansion joints are subjected to stresses, they undergo expansion and compression strains. These can be axial, transverse or angular (Figure 7.3) and are explained below.

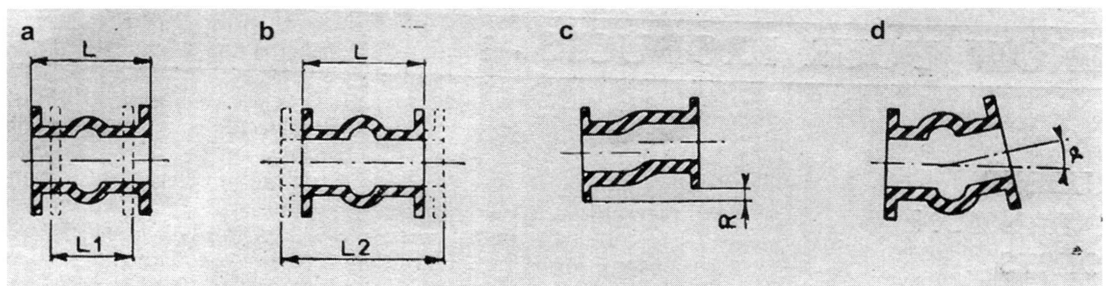

Figure 7.3: (a) Axial compression; (b) axial elongation; (c) transverse deflection; (d) angular deflection

Axial compression/extension: axial compression is dimensional shortening of the expansion joint along its longitudinal axis, whilst axial extension is lengthening along the same axis.

Transverse deflection: the relative displacement of the two ends of an expansion joint perpendicular to its longitudinal axis is called transverse deflection.

Angular deflection: the displacement of the longitudinal axis of the joint into a circular arc is known as angular deflection.

As is known, rubber differs from steel in that it can accommodate large deformations when stressed, and its recovery is rapid and complete. Including arches or bends in expansion joint design takes advantage of the elasticity of rubber, and so increases its capacity to accommodate movement.

Constructional features

The fabrics used for reinforcement of the expansion joints are Nylon, Rayon cord or Cotton Duck. They are resin coated and dried before use. Steel cord or thin steel plates are also used as plies when resistance to fatigue failures and maximum strength are required. A rubber expansion joint without reinforcement will tend to have larger elongation and compression, making the joint more flexible and unstable during operation. Neoprene-based adhesive solution is used for bonding the fabric with its substrate. The flanges are integrally inbuilt and molded. The constructional parts of the rubber joints are the carcass or body, the cover, the inner tube, and the flanges (Figure 7.4).

Carcass or body: the part other than the flanges is called the carcass, or body. This consists of the reinforcement, inner tube and outer cover.

Cover: the cover of the rubber expansion joint is made out of a rubber which resists weathering attack and mechanical damage, if any, caused during handling and installation. In general,

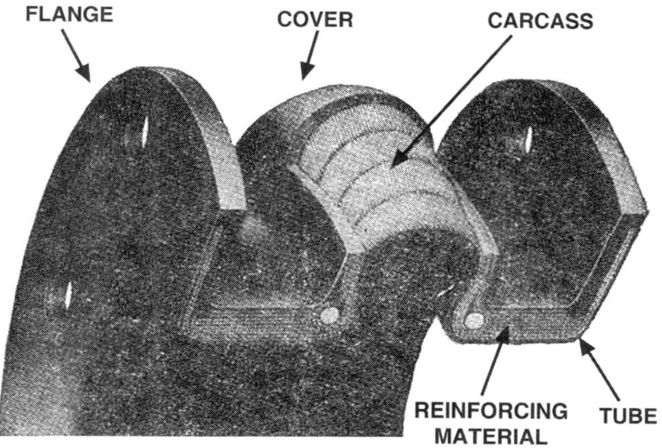

Figure 7.4: A typical expansion joint

neoprene, natural rubber, butyl rubber, nitrile rubber, EPDM rubber, styrene butadiene rubber or Hypalon are used here.

Flanges: flanges have thicker sections than the wall of the joint and general dimensional specifications as per piping standards.

The tube: the tube in an expansion joint can be considered as a lining material which is integrally bonded with the body of the joint. The tube extends up to the edge of the flanges and it protects the body and the carcass against the possibility of attack by the fluid handled, if a leak occurs.

The temperature ratings and application limits of certain rubbers used in the manufacture of rubber expansion joints are given in Table 7.1.

TABLE 7.1 Temperature ratings and application limits of rubbers used in rubber expansion joints

Rubbers	Temperature (°C)	Applications
Natural rubber	−30−+100	Resistant to most organic chemicals, acids etc
Neoprene rubber	−20−+120	Resistant to flame, ozone attack, halogenated and aromatic hydrocarbons
Chlorosulfonated polyethylene	−5−+125	Resistant to aliphatic hydrocarbons, corrosive and oxidizing fluids, sodium hypochlorite and sulfuric acid
Butyl rubber	−20−+120	Resistant to weathering and corrosive acids and chemicals
Nitrile rubber	−20−+100	Oil and fuel resistant, not affected by dilute alkali solutions, aliphatic hydrocarbons etc
Viton	−20−+200	Resistant to mineral acids and alkalis
Silicon	−60−+320	Resistant to weathering, solvents and chemicals at high temperatures where other rubbers will fail

MANUFACTURE OF RUBBER EXPANSION JOINTS

In the manufacture of rubber expansion joints, material selection and formulation should be considered against the constructional characteristics required in the product. Rubbers, compounds, adhesives and solvents are all thoroughly checked for quality, to avoid possible layer separation or delamination of the layers while the joints are being built.

Rubber expansion joints are compression molded in hydraulic presses. While constructing large joints, split type forms that are made either of cast iron or mild steel, are employed. Rubber and coated fabric layers are built up on the forms using suitable adhesives and solvents. After building, the joint is vulcanized in an autoclave under about $4\,kg/cm^2$ steam pressure for about 4–5 h. After vulcanization, the product is then stripped from the form. In general, the tube and the flanges of the joints are thicker than the arch, since the tube is required to bear load. The arch is thinner and, because of its shape, it acts as the bellows which protect the tube from forces exerted by movement. Where more flexibility is required, a double arch is used. In this case, the face-to-face length of the expansion joint is larger than when a single arch is designed.

When minimum movement capability is required, the arch is sometimes filled with soft rubber using a suitable adhesive. The maximum amount of movement (axial extension and compression, lateral deflection and angular rotation) that an expansion joint is capable of absorbing is called the rated movement. This rating depends on various factors, such as the size of the expansion joints, the thickness of the tube, arch or convolution, and the type and properties of rubber compound and fabric used in construction. Rated movements are established by manufacturers of expansion joints theoretically, or are based on actual load deflection curves of each size of joint. Rubber expansion joints are generally subjected to hydraulic and vacuum tests at 1.5 times the operating pressure. No internationally accepted standard technical specification for rubber expansion bellows is available, since they are mostly custom built to specific operational requirements. The Expansion Joint Manufacturers Association in New York has laid down standards for rubber expansion joints, which are called EJMA standards [2].

REFERENCES

1 www.hse.gov.uk/nuclear/thorp.htm
2 Handman, S.E. Piping systems. In Kirk-Othmer's encyclopedia of chemical technology, 4th edn, (Kroschwitz, J.I., Howe-Grant, M., eds), Vol. 19, pp 906–907. Wiley-Interscience, New York, 1992

CHAPTER 8

SWELLING ASPECTS OF RUBBER RELATED TO SEAL PERFORMANCE

There has been a long-standing commercial need for elastomers with a high degree of oil resistance. Such materials are very important to many industries where components come into contact with fluids (hydrocarbon and otherwise). Engine seals, specifically, should be able to withstand the fluids they are sealing without either becoming too soft or too brittle.

Major synthetic elastomers used commercially are indicated below:

- Oil-resistant

  o Oneoprene

  o nitrile rubber

  o ethylene–acrylate elastomers

  o polyacrylate elastomers

  o Hypalon

  o polyurethane elastomers

  o silicon rubber

  o fluorocarbon elastomers

- Non-oil-resistant

  o ethylene–propylene rubbers (EPM, EPDM)

  o natural rubber

  o butyl rubber

  o styrene–butadiene rubber

  o polybutadiene rubbers

  o polyisoprene rubbers

Rubber Seals for Fluid and Hydraulic Systems 2010; ISBN: 9780815520757

- Miscellaneous rubbers

  - thermoplastic rubbers

  - polysulfide rubbers.

While the focus is on oil-resistant rubbers, there is also interest in materials that display enhanced heat resistance.

Swelling is a serious drawback for rubber products, such as seals and 'O' rings, which are used in contact with liquids. This can be reduced to some extent by suitable formulation, but a much better solution to the problem is offered by specialty synthetic rubbers, such as neoprenes, nitriles and thiokols. In addition to actual swelling, fluids also affect the mechanical properties of the rubber, as discussed earlier. The swollen rubber becomes slightly less extensible and much weaker, for example, the tensile strength of a pure gum compound that has been swelled in gasoline for 7 days is reduced from 2800 psi to 215 psi, and its elongation at break from 700% to 500%. Its hardness is also significantly reduced.

VOLUME CHANGE

In most cases, swelling takes place as liquid is adsorbed. The diffusion rate of liquid in to a rubber test piece, and also its size will determine the time taken to reach equilibrium. After this the rate of absorption of liquid slows. The lower the viscosity of the liquid, the higher the diffusion rate [1]. When a rubber is immersed in a contacts a fluid, its surface can be attacked chemically. Initially this is negligible or even nil in the case of chemically inert rubbers. Swelling does take place its depending upon the solubility characteristics of the rubber and the fluid. With swelling, the rubber is strained and however, the amount physical properties change. Prolonged contact with fluids leads to leaching, or extraction of certain chemicals, especially antioxidants, which has a significant influence on seal life. The main change noticed after immersion of a rubber in a fluid for a specified period is the change in volume. The various types of rubbers vary quite substantially in their permeability to fluids. This is an intrinsic property, which can only be altered marginally by compounding.

The higher the temperature, the greater the swelling. If there is no chemical change taking place, the phenomenon is reversible as, for example, a silicone rubber, which can be swollen and shrunk repeatedly by hot and cold ester lubricants. The high diffusion rate in silicone rubber enables this to occur in a short time cycle of (24 h).

Practically, the rubber in a seal is under strain nearly all the time, so access of fluid is restricted by its enclosure in a housing or groove. In addition to this, the volume of the groove may restrict swelling to levels below its theoretical maximum. If the diffusion rate is low, fluid absorption will not immediately increase the seal's volume. If the diffusion rate is high, and the equilibrium swelling very temperature dependent, shrinking can occur during the lowering stage of the normal temperature cycle to most equipment. During hot immersion of a silicone rubber, the volume change increases with temperature. At this stage, degradation is inevitable. A silicone rubber seal will generally vary its volume with short term temperature changes, whereas a fluorocarbon rubber seal will steadily increase its volume over the longer term [2]. Shrinkage of a rubber seal, by evaporation is also important in seal function.

SWELLING UNDER STRAIN

The volume of liquid absorbed by a rubber vulcanizate is altered by the application of a stress, being increased by stretching and decreased by compression. The latter is more relevant in sealing applications. A housing can only accommodate a certain degree of swelling before it becomes overfull, leading to extrusion and potential damage to the seal. Practice in dealing with this varies, although a volume increase of 15% is usually unacceptable. Shrinkage of the rubber can occur under certain circumstances, due to the extraction of an ingredient in the rubber by the liquid. This is unacceptable at levels above 1 or 2%. While, tensile strength or elongation at break do not necessarily affect seal performance, they can be used to follow changes taking place as temperature is increased in the presence of fluid. Hardness changes due to swelling are of practical importance, however. If too much softening takes place, the seal may yield under pressure. On other hand if too much hardening takes place, the seal may not absorb movement.

In addition to hardness, flexibility is often evaluated by a bend test.

The swelling of neoprene vulcanizates by oils and solvents may be considerably reduced by increasing the state of cure, increasing the amount of filler loading in the compound, or by using a plasticizer miscible with the immersion fluid and consequently extracted by it. Neoprene type WHV compounds, when highly extended, have excellent resistance to swelling because of the low polymer content in the formulation. Vulcanizates containing fine carbon black swell less in petroleum fluids than those containing an equal volume of coarser carbon blacks. The results of a study carried out by D.C. Thompson [3] on the volume increase of neoprene rubber in various fluids are given in Table 8.1. This is useful as a general guide and reference while designing seal compounds and adhesive formulations.

SWELLING TESTS

Swelling of rubbers by fluids is known to be a diffusion-controlled process and, up to the equilibrium swelling, the volume of liquid absorbed is proportional to the square root of the time that the rubber has been immersed in the liquid. For most liquids, this rate also depends on its the viscosity, rather than its chemical nature.

In coping with swelling, there is no universally acceptable standard for determining the suitability of a rubber seal for an application. The general practice is to immerse a vulcanized

TABLE 8.1 Volume increase (%) after 7 days immersion at 25°C (77°F)

Fluid	SRF carbon black	
	0	50 phr
Acetone	55	40
Aniline	110	75
Benzaldehyde	365	140
Butter	5	5
Butyric acid	80	40
Carbon tetrachloride	350	170
Cotton seed oil	5	5
Cyclohexane	119	75
Cyclohexanone	370	185
Dibutyl phthalate	200	110
Ethyl acetate	95	60
Ethyl alcohol	10	0
Ethylene glycol	0	0
Furfural	40	25
Isopropyl ether	80	35
Jet fuel JP4	60	45
Kerosene	60	
Linseed oil	20	10
Methylene chloride	335	150
Methyl ethyl ketone	170	85
Nitrobenzene	300	130
Oleic acid	80	40
Olive oil	5	5
Orthodichlorobenzene	500	375
Tetraethyl lead	75	35
Toluene	415	185
Trichloroethylene	450	190
Turpentine	215	100

rubber sample in various fluids and record a range of parameters, such as change in weight volume thickness, and hardness. These data can be yardsticks for approximate assessment of a seal's suitability. Sometimes users of rubber products specify their own test requirements in terms of either weight or volume increase under given conditions.

The suitability of a rubber for use in contact with an oil or chemical is thus assessed by immersing a sample in fluid and measuring the resulting increase in weight or volume. Such tests represent more severe conditions than are generally encountered in actual service, since few commercial seals are immersed for longer than 24 h and, even then, only partial immersion is likely. In actual use, only a part of the seal's surface area is exposed to fluid attack. Tests have shown that totally immersed test specimens swell three to six times more than those exposed to fluid on one surface only. It makes sense therefore that a stock may perform well in use, even though in total immersion tests the volume swell may be 100% or more. On the other hand, the fact that a compound is found satisfactory for a given application is no assurance that the same compound will be found suitable under different conditions. When determining whether a seal made from a given neoprene rubber will be satisfactory for a given application, the seals should be tested under the actual service conditions. For aircraft seals, approval by the concerned directorates of military and civil aviation are given as 'type approval'. This is only given after rig testing under simulated conditions where even small leaks are not acceptable. Many companies such as M/s Haughton Design Limited (Engineering Design Consultants, Staffordshire Technology Park, Beaconside, Stafford, UK) provide rig testing solutions for critical components such as seals, and the components in which they are fixed.

If such test rig programs are impractical due to economic reasons, it is better to devise a test which simulates actual service conditions as closely as possible, rather than rely entirely on more theoretical test results.

EFFECT OF TEMPERATURE

It is common practice to carry out immersion test at various temperatures up to and above the stability limits of the fluid. In the case of volatile liquids, these tests are carried out at elevated pressures. The effect of oxygen should not be overlooked in laboratory tests, particularly when, in actual application, air is excluded. Oxidation or thermal breakdown may produce decomposition products which are more reactive than the original liquid. Chemical changes can substantially modify the simple physical rubber/fluid equilibrium and this becomes more marked as the temperature increases and if esters rather than simple hydrocarbons are used as fluids. The cycle of operation in service should be carefully considered when interpreting laboratory results. A thick section of rubber, a low diffusion rate of the liquid into the rubber, and a short, high temperature cycle will lead to a much lower equilibrium or maximum swelling and to lower chemical attack than a thin rubber section, high diffusion rate and continuous operation at high temperature.

Yet another two effects should also be considered:

1. Confining of the seal within the housing

2. Differences between the coefficient of thermal expansion of the rubber and that of the metal housing. Silicones generally have higher values than other rubbers.

As temperatures increase, so will deflection or strain on the seal, but even so, it is expected that its uptake of liquid will be less when it is in a housing, since less of its surface area is exposed. Correspondingly, a fall in temperature will result in a reduction of seal size relative to the groove as is obvious when leaks occur.

Whatever may be the case, regardless of any stress imposed on the rubber, it must recover its shape to seal the gap and prevent leaks. If temperature and pressure are kept constant and no disturbance of the metal/seal contact occurs, a seal will continue to function for a considerable time.

ADVANTAGES AND LIMITATIONS OF VARIOUS RUBBERS

Some of the advantages and limitations of neoprene rubbers are given in Table 8.2, and those for nitrile rubber in Table 8.3.

Nitrile and neoprene are high volume oil-resistant elastomers. Nitrile is superior to neoprene in resistance to oil, gasoline and aromatic solvents. However, it does not perform as well as neoprene in applications requiring exposure to weather, ozone and sunlight. Furthermore, it does not have inherent flame resistance.

Some of the advantages and limitations of the following elastomers are given in the text: ethylene acrylate, Table 8.4; acrylate, Table 8.5; Hypalon, Table 8.6; chlorinated polyethylene, Table 8.7; polyurethane, Table 8.8; silicone rubber, Table 8.9.

TABLE 8.2 Advantages and limitations of neoprene rubbers

Advantages	Limitations
Good inherent flame resistance	Poor or fair resistance to aromatic and oxygenated solvents
Resistance to oil and gasoline	Limited flexibility at low temperatures
Very good resistance to weather, ozone and natural aging	
Good resistance to abrasion and flex cracking	
Very good resistance to alkalis and acids	

TABLE 8.3 Advantages and limitations of nitrile rubber

Advantages	Limitations
Very good resistance to oil and gasoline	Inferior resistance to ozone, sunlight and natural aging
Very good resistance to alkalis and acids	Poor resistance to oxygenated solvents

TABLE 8.4 Advantages and limitations of ethylene acrylate elastomers

Advantages	Limitations
Excellent resistance to heat, ozone and sunlight	Poor resistance to aromatic and oxygenated solvents
Moderate resistance to oil, gasoline and hydraulic fluids	Limited flexibility at low temperatures
High energy absorption (damping)	Inferior resistance to alcohols and acids

TABLE 8.5   Advantages and limitations of acrylate elastomers

Advantages	Limitations
Outstanding resistance to heat and hot oil	Poor resistance to alcohols and alkalis
Excellent resistance to weather, ozone, sunlight and oxidation	Limited flexibility at low temperatures
Very good resistance to oil and gasoline	Inferior resistance to water and steam

TABLE 8.6   Advantages and limitations of Hypalon rubbers

Advantages	Limitations
Good flame retardant properties	Poor or fair resistance to aromatic solvents
Moderate resistance to oil and gasoline	Limited flexibility at low temperatures
Superior resistance to weather, ozone, sunlight and oxidation	
Good abrasion resistance	
Excellent resistance to alkalis and acids	

TABLE 8.7   Advantages and limitations of chlorinated polyethylene

Advantages	Limitations
High tear/tensile strength	High gas permeability, sensitive to pH, must be stabilized
Good abrasion resistance	Generation of hydrochloric acid vapor upon combustion
Excellent resistance to weather, ozone and ultraviolet rays	
Good resistance to oil, gasoline, chemicals	
Good low temperature properties	
Flame retardancy	

TABLE 8.8   Advantages and limitations of polyurethane

Advantages	Limitations
Outstanding resistance to abrasion and tear	Poor resistance to alkalis, acids and oxygenated solvents
Very high tensile strength with good elongation	Inferior resistance to hot water
Excellent resistance to weather, ozone, and sunlight	
Good resistance to oil and gasoline	
Excellent adhesion to fabrics and metals	

TABLE 8.9 Advantages and limitations of silicone rubber

Advantages	Limitations
Outstanding resistance to high heat	Poor resistance to abrasion, tear and cut growth
Excellent flexibility at low temperatures	Low tensile strength
Low compression set	Poor resilience
Very good electrical insulation	Inferior resistance to oil, gasoline and solvents
Excellent resistance to weather, ozone, sunlight, and oxidation	Poor resistance to alkalis and acids
Superior color stability	

Hypalon is a close match to neoprene in most properties, but it is superior in resistance to acids, solvent, ozone and oxidation, and has decidedly better color stability. This rubber can be used instead of polychloroprene or plasticized polyvinyl chloride in many applications.

Dimethyl silicone rubbers show good performance in acetone and diesters, but undergo up to 200% swell in aliphatic and aromatic hydrocarbons. The replacement of one methyl group on each silicon atom with a polar group such as trifluoropropyl reduces the swelling in aliphatic and aromatic hydrocarbons to less than 25%, but increases the swelling in acetone and diesters.

Tables showing advantages and limitations of these elastomers follow in the text: fluorocarbons, Table 8.10; EPDM, Table 8.11; natural rubber, Table 8.12; butyl rubber, Table 8.13; styrene butadiene, Table 8.14; polybutadiene, Table 8.15; and polyisoprene, Table 8.16.

TABLE 8.10 Advantages and limitations of fluorocarbon elastomers

Advantages	Limitations
Outstanding resistance to high heat	Poor resistance to tear and cut growth
Excellent resistance to oil, gasoline, hydraulic fluids and hydrocarbon solvents	Very little resistance to oxygenated solvents
Very good resistance to weather, ozone and sunlight	Fair adhesion to fabric metals
Good flame retardant	
Very good impermeability to gases and vapor	

TABLE 8.11 Advantages and limitations of EPDM polymers

Advantages	Limitations
Excellent resistance to heat, ozone and sunlight	Poor resistance to oil, gasoline and hydrocarbon solvents
Very good flexibility at low temperatures	Adhesion to fabrics and metal is poor
Good resistance to alkalis, acids and oxygenated solvents	
Excellent color stability	
Superior resistance to water and steam	

TABLE 8.12 Advantages and limitations of natural rubber

Advantages	Limitations
Outstanding resilience	Poor resistance to heat, ozone and sunlight
High tensile strength	Little resistance to oil, gasoline and hydrocarbon solvents
Superior resistance to tear and abrasion	
Excellent rebound elasticity	
Good flexibility at low temperatures	
Excellent tack, self-adhesion	
Excellent adhesion to fabrics and metals	

TABLE 8.13 Advantages and limitations of butyl rubber

Advantages	Limitations
Outstanding impermeability to gases and vapor	Poor resistance to oil and hydrocarbon solvents
Very good resistance to heat, oxygen, ozone and sunlight	Fair processability
High energy absorption (damping)	Poor resilience
Excellent resistance to alkalis and oxygenated solvents	Low rebound elasticity
Good hot tear strength	
Superior resistance to water and steam	

TABLE 8.14 Advantages and limitations of styrene butadiene rubber

Advantages	Limitations
Excellent impact strength	Poor resistance to ozone and sunlight
Very good resilience, tensile strength, abrasion and flexibility at low temperatures	Very little resistance to oil, gasoline and hydrocarbon solvents

TABLE 8.15 Advantages and limitations of polybutadiene rubber

Advantages	Limitations
Outstanding resilience	Inferior processability
Excellent flexibility at low temperatures	Poor resistance to oil, gasoline and hydrocarbon solvents
Superior resistance to abrasion, cut growth and flex cracking	Very little resistance to heat and ozone

TABLE 8.16 Advantages and limitations of polyisoprene rubber

Advantages	Limitations
Outstanding resilience	Poor resistance to heat, ozone and sunlight
Superior resistance to tear and abrasion	Very little resistance to oil, gasoline and hydrocarbon solvents
Very good tensile strength	
Excellent rebound elasticity	
Good flexibility at low temperatures	
Lack of odor	

TABLE 8.17 The glass-transition temperatures of various rubbers

Rubbers	Tg (°C)
Cis-1,4-polybutadiene	−105
Cis-1,4-polyisoprene	−70
Polyisobutylene	−70
Polybutadiene (emulsion polymerized; 20% 1, 2)	−85
Polybutadiene (sodium catalyzed; 60% 1, 2)	−46
Ethylene-propylene copolymer (50/50)	−60
Poly(butadiene-co-styrene) 77/23	−56
Poly(butadiene-co-styrene) 64/36	−38
Poly(butadiene-co-acrylonitrile) 80/20	−56
Poly(butadiene-co-acrylonitrile) 70/30	−41
Poly(ethylacrylate)	−22
Poly(n-butylacrylate)	−56
Poly(n-octyl methacrylate)	−20
Polyesterurethane	−35−−50
Polychloroprene (85% trans-1, 4)	−45
Poly(perfluoroproypylene-co-vinylidene fluoride)	−55
Polydiemthylsiloxane	−120

TABLE 8.18 Solubility parameters for common rubbers

Rubbers	δ (cal/cm^3)$^{1/2}$
Polytetrafluoroethylene	6.2
Polyisobutylene	7.85
Polyethylene	7.9
Polybutadiene	8.4
Polystyrene	9.1
Poly (methyl methacrylate)	9.45
Poly (vinylchloride)	9.6
Poly (ethyleneoxide)	9.9
Nylon 6.6	13.6
Cellulose triacetate	13.6

The glass-transition temperatures of various rubbers are shown in Table 8.17, and the solubility parameters of certain rubbers and solvents in Tables 8.18 and 8.19 for reference.

SIDE-CHAIN GROUP VERSUS OIL RESISTANCE

As the electronegativity of the side-chain group increases on a rubber, so does its oil resistance. Polyisoprene, with only a hydrocarbon side group, has almost no resistance at all. Polychloroprene is polyisoprene with a chlorine atom replacing the methyl group side chain. This chlorine atom is more electronegative than the methyl group and so increases the oil resistance

TABLE 8.19 Solubility parameters for some common solvents

Solvent	δ (cal/cm^3)$^{1/2}$
Difluorodichloromethane	5.1
n-Decane	6.6
Dibutyl amine	8.1
Cyclohexene	8.2
n-Butyl acetate	8.3
Carbon tetrachloride	8.6
Xylene	8.8
Toluene	8.9
Benzene	9.2
Styrene	9.3
Acetone	9.9
Methanol	14.5
Water	23.4

TABLE 8.20 Structures versus oil, swell relationship of rubbers

Polymer	#3 Oil swell (%)
Fluorocarbon	10
Nitrile (33%)	40
Polyacrylate	60
Polychloroprene	120
Polyisoprene	150

TABLE 8.21 Percent swelling of synthetic rubbers by various solvents

Solvent	Benzene	Heptane	CCl$_4$	Ethanol	MEK	Acetone	H$_2$O
Polybutadiene	$\approx$500	$\approx$500	300–500	$\approx$1	50–70	3–10	$\approx$1
Polyisobutylene	$\approx$500	$\approx$500	300–500	$\approx$1	50–70	3–10	$\approx$1
Cis-polyisoprene	$\approx$500	$\approx$500	300--500	$\approx$1	50–70	3–10	$\approx$1
Polybutadiene-co-styrene(76/24)	$\approx$500	$\approx$500	300--500	$\approx$~1	50–70	3–10	$\approx$1
Chlorosulfonated polyethylene	>200	<10	.150	$\approx$0	100	20	$\approx$0
Polychloroprene	200	25–35	230	1	70	30	$\approx$0
Styrene-acrylonitrile (85/15)	200	70	>200	10–20	100	80	–
Viton A	20	1	1	2	290	280	3

of the material. Polyacrylate and acrylonitrile have ester and cyanide side groups, respectively, which are also electronegative and thus provide oil resistance. Fluorocarbons have many fluorine side groups that are even more electronegative and provide very high oil resistance (Tables 8.20 and 8.21).

Oil-resistant synthetic rubbers and their polymerization type are given in Table 8.22.

TABLE 8.22 Oil-resistant synthetic rubbers and their polymerization type

Rubber	Tg (°C)	Polymerization
Neoprene (polychloroprene)	−45	Free-radical (emulsion)
Nitrile rubber (acrylonitrile-butadiene)	−41--−56	Free-radical (emulsion)
Ethylene-acrylate elastomers	−35	Free-radical (emulsion)
Polyacrylate elastomers	−20--−56	Free-radical (emulsion, suspension)
Hypalon (chlorosulfonated polyethylene)	0--−30	Polymer modification
Chlorinated polyethylene	0--−30	Polymer modification
Polyurethane elastomers	−35--−50	Step growth
Silicone rubber	−120	Ring–opening
Fluorocarbon	−55	Free-radical

REFERENCES

1 Blow. C.M. Properties of rubbers for sealing application. Unclassified report no 142, Ministry of Technology, Directorate of Material Research and development, May, 1967

2 Blow, C.M., Exley, K., Southwart, D.W. The penetration of liquids in rubbers. Journal of the IRI. 3(1), 1969

3 Thompson, D.C. Report 57-1 Practical neoprene compounds of low elastomer content. Rubber Age 72, 17 1953; quoted in The neoprenes, p. 76. E I Dupont De Nemours & Co Inc., Wilmington, Delaware, 1964

CHAPTER 9

RUBBER TO METAL BONDING

BACKGROUND

Rubber to metal bonding is a general phrase covering a number of interdependent processes. The units resulting from the process are used as metal bonded oil seals, engine mounts for the isolation of noise and vibration in automotive and engineering applications, rubber lined mild steel tanks, and chemical process equipment for corrosion prevention. Larger units are used to decouple translational movement for bridges and buildings. The technology allows the production of a uniform, high quality product that is free from failure.

Although the birth of the rubber industry was over 150 years ago, it is during the last 75 years that design engineers have been able to combine the strength of metals with the elasticity of rubber. Rubber bonding used to be done by mechanical means. Probably the first real bond between rubber and metal was achieved through the hard rubber technique [1]. The first commercially successful bonding process was achieved by the brass plating technique during the 1920s. In this process, the ratio of copper to zinc in the brass was maintained at a specific level to achieve proper bonding. Compounding of rubbers for bonding with the brass plating technique had severe limitations and this technique has become outdated. By the end of the Second World War, bonding with chemical agents began gradually to replace earlier processes. The hard rubber or ebonite based bonding system is still followed in some applications, such as rubber rollers and tank linings.

In the chemical bonding technique, three essential elements form the core of the process; the rubber compound, the bonding agents and the substrate. Selection of the polymer base and associated compound depends mainly on the product specification. Provided that the rubber can flow into the mold without developing significant levels of cross-linking (less than 2%), a bond can be formed from any rubber compound. There are no restrictions on compounding ingredients, although it is best to avoid substances that will bloom rapidly on the surface of the uncured stock. The rubber chemist, therefore, needs to concentrate his attention to matching the physical requirements of the cured rubber–metal product to the processing needs of the molding process, and other forming processes. It is generally agreed that that bonding of rubber with metal occurs through

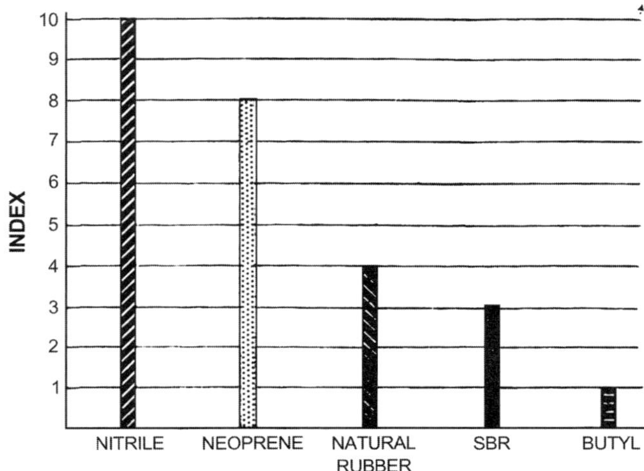

Figure 9.1: Bondability index of common elastomers

both physical and chemical means. Since metals are polar, polar rubbers will be more easily bound than non-polar ones. From this, the concept of a bondability index has been developed [2,3]. Figure 9.1 shows the bondability index of some of the common rubbers. The higher the index, the more readily the rubber bonds with metals when using a single coat adhesive system. The lower the index, the greater is the necessity for using a two-coat system – a primer to give good bonding with metal, and a secondary or top adhesive coat to give good bonding with rubber. Many bonding agents are available, the most important ones being:

1. polyisocyanates

2. chlorinated rubber

3. phenol formaldehyde resins.

Organic polyisocyanates, such as triphenyl-methyl-tri-p-isocyanate, are useful for bonding natural rubbers to mild steel, cast iron substrates, and light alloys. A solution in methylene chloride is sold as Desmodur-R by M/s Bayers, Germany. Other brands are also available. Chemlok brand adhesives, based on isocyanates, are manufactured by M/S. Lord Corporation, USA and widely used, although there is a tendency for the isocyanate adhesive to be squeezed out during the molding process under hydraulic pressure. This can be partially avoided by adding chlorinated rubber to the solution.

Chlorinated rubber is also an effective bonding agent. It can be used for bonding neoprene, nitrile, and natural rubbers to metals. Phenol formaldehyde resins have been used alone, or in conjunction with chlorinated rubbers, but curing time is lengthy.

ELEMENTS OF THE RUBBER/METAL BOND

The three important elements of the rubber/metal bond are:

1. the bonding layer

2. selection of the bonding agent

3. choice of the substrate.

For many years, bonding agents were a primer coat based on phenolic-style resin, and a top-coat formed from solutions of polymers and other ingredients. The formulations of these materials were not disclosed by rubber manufacturers. Bond formation appears to be associated with the development of a very high modulus in the rubber layer immediately adjacent to the surface of the substrate.

The selection of bonding agents depends on the type of rubber to be bonded, the hardness/modulus of the rubber and the design of the component. The selection process is critical to the robustness of the bond. Water-based versions of bonding agents have been introduced which, after much development, are now effective replacements for the solvent-based agents. Bonds tend to be up to 10% weaker, but components show good resistance to normal conditions found in automotive applications. Application methods for water-based bonding agents are similar to those for the solvent systems, but inserts do need to be preheated to 60–80°C before application of the primer, and then reheated before applying the topcoat. Drying times are surprisingly short and are no barrier to high volume production.

The choice of substrate rests solely with the component designer, who must consider the necessary strength and durability requirements for use. The traditional substrate is steel, in all its forms and grades, but increasing use is made of aluminum alloys and polyamides to save weight. Almost any material can be bonded to rubber, provided that it can withstand the heat and pressures of the molding process, which for practical purposes, eliminates polyolefin plastics. Polyacetal inserts can be bonded, but require careful etching and use of rubber molding temperatures below 150°C. Polytetrafluoroethylene (PTFE) provides a useful low friction material for use in anti-roll bar bushes as well as fluid seals. It can be bonded successfully to rubber by chemically etching the surface of the plastic prior to application of bonding agents. Its use in such applications has, however, been largely superseded by woven PTFE/Terylene fiber material, which offers a better, mechanical bond.

THE BONDING PROCESS

Substrate cleaning

The first key step in the preparative process for metal inserts is to clean them. To prepare steels, all traces of oil, grease or solid lubricant must be completely removed from the metal surface.

For this, degreasing and shot blasting are widely employed, although wet blasting followed by a phosphate conversion treatment is gaining greater acceptance as a cost-effective process, and it also gives the added benefit of improved corrosion resistance. Whatever process route is taken, the emphasis should be on control, to ensure consistency of results. Degreaser plants need to be regularly monitored to minimize any build up of contaminants, or change in pH. In practice, use of neutral stabilized trichloroethylene is suitable for vapor degreasing. The grade of grit used in shot-blast operations is important, and will affect the environmental resistance of the finished product. The profiling of particle sizes is a useful tool to monitor machine effectiveness, and ensures that dust levels remain low during blasting. Routine sampling will show if fresh grit is being fed to the machine and so detect a failure of the dust extraction mechanism. Differences in surface profile due to changes in the size range of the grit will not cause bond failure, but the presence of dust and debris in the grit will. Dust adheres to the newly cleaned metal surfaces and is very difficult to remove, even with a second degreasing operation.

Application of bonding agents

Methods for applying bonding agents are subject to frequent review. The process requires the application of a primer coat and then a topcoat. In high volume production methods, problems with blocked guns, or other mechanistic faults, may allow components to be produced without primer below the topcoat. Such parts will initially show a good bond, but will be likely to fail in service. The best method to date is to ensure that the primer is sprayed over a slightly wider area than the topcoat to show that it is there. However, this does not work for inserts that need to be bonded over 100% of their surface. If service conditions are exceptionally severe, as is the case for bonds which are subject to dynamic strain, or in contact with oils, solvents, boiling water or corrosive chemicals, then a two-coat adhesive system is preferred. When using a two-coat system, it is essential that there should be good compatibility between coats. Lack of compatibility is likely if primer and top or cover coat are obtained from different suppliers.

Environmental awareness has prompted manufacturers to find means of reducing or eliminating the volume of solvent used in cleaning and coating processes. The introduction of low pressure, high volume (LPHV) spray guns has reduced the volume of bonding agent used by 20%.

Rubber molding

The rubber molding operation brings together all three elements of the process mentioned above, and so is the most critical point in the process. If one factor in the production cycle for the inserts and bonding agent is wrong, then the product will fail. For automotive production, the preferred method of molding is by injection. This allows the greatest control over the process, allowing the condition of the rubber as it enters the cavity to be tailored to give optimum product quality. Satisfactory bonding requires adequate molding pressure during the curing cycle. Molds should fit well to allow adequate venting.

Mold design

Molds need to be designed to ensure exact balance between cavities and the elimination of trapped gases. The presence of gases causes a high incidence of bond failures through the 'diesel effect' (this a reference to the combustion process in a diesel engine), where elements of the bonding agent film burn under the combined effects of heat and high-pressure gas. Lack of balance between cavities will result in some imperfectly formed components with bonds that may fail. Problems such as these are avoided by study of rubber flow. The mold consists of two parts for 'O' rings – a base plate with locating pins, and a top plate. Single or multicavity molds are made according to production requirements, and the platen size of the hydraulic molding press. The molds for metal bonded seals are generally of three parts – a fixed central core mating with the top plate of the mold, and a center ring which holds the metal shell. The molds are made with mild steel or 'Nitralloy', and case hardened. A cyclic molding process is adopted for increased productivity when the number of molds per platen is small and multidaylight presses are employed.

COMPOUNDING OF RUBBER

Compounding is important for achieving a good bond. It has previously been overlooked, when focus was only on adhesive systems and substrate profiles. Very often, small changes in the formulation of a compound increase the level of bonding, particularly the types and dosages of antioxidants to fillers, either inert or reinforcing.

In order to attain strong bonds, it is essential to ensure 100% wetting of the compounded stock by the adhesive cement. Losses of the active ingredients during mixing should be avoided to enable primary chemical bonds to be formed. This means that at the time when the rubber–metal interface is formed there should be no sign of vulcanization. The curing system should be designed to give the optimum delayed action that is consistent with maximum productivity. Mixing conditions should therefore be specified to ensure that this delayed action is consistent from batch to batch. A curometer trace could be obtained on each batch at the curing temperature as a quality control check.

In general, the higher the filler content, the stronger the rubber–metal bond, carbon black is the best filler; wherever possible, at least 50 phr of carbon black is suggested for all compounds, with suitable adjustments of dosage levels of other ingredients, such as process aids etc. In rubbers with a high bondabilty index, all kinds of carbon blacks can work well. However, channel blacks are preferred for those with a low bondability index. It is very important that softening or processing oils are chosen with extreme care. If the oil is not fully compatible with the rubber, bonding will be severely affected, due to leaching of the oil from the rubber–metal interface. Aromatic and ester type oils are generally unsatisfactory, whilst best results are obtained with naphthenic oils. Highly oil extended synthetic rubbers are not recommended for bonded products. Blooming ingredients should be avoided, or their effect reduced by keeping dosages low. Pre-warming of the stock before molding is a good practice to get a fresh bondable rubber surface during molding. It is worth repeating that the most important aspect of compounding is to curing sufficiently delay; to ensure that it does not start until the rubber to bonding agent interface is formed.

Since most bonding agents are dispersions of insoluble ingredients, a homogeneous dispersion can be made by stirring while applying, in order to ensure uniform spread of the ingredients in the adhesive film. This is even more important if make-up diluents are used, which can lead to settling of the dispersed ingredients. Viscosity and solid content should be checked regularly, as a quality control measure.

The dipping method for adhesive coating of metal disks is best for oil seal manufacture. Viscosity and solid content of the dipping bath is maintained by slow and constant agitation, both vertically and horizontally. These variables should be checked regularly, as part of quality control procedures.

For manufacturing reinforced inflatable seals, fabric is reinforced by dipping in suitable adhesive. Aqueous synthetic latex and resorcinol formaldehyde adhesive solution is used as a dipping solution for treating the rubber sheets that will be used for the cutting blanks that will be fed into the molds. The adhesive is applied by passing the woven fabric through a dip tank and controlling the adhesive pick-up. It is then dried in an oven before applying the rubber sheet to the coated fabric, which is usually either cotton or Nylon.

REJECTIONS IN MOLDING

Even in ideal conditions, there will be rejections during the molding process, but this can be kept to a minimum needs to be carried out. If the rejection rate is more than the accepted level of around 5%, a systematic evaluation of the reasons for failures. Observation of the rejected metal bonded seals shows that, four types of failures can be observed.

The first is an adhesive–metal bonding failure resulting in clean metal showing either in parts or over the whole of the bonded area, which is not related to the rubber.

The second failure is in the adhesive–primer coat, which can be easiliy identified, and is invariably due to application of the top adhesive coat before the primer has dried out properly.

The third is rubber–top coat failure, indicating separation of the rubber from the bonding agent, and is shown by hard and relatively fresh, shining areas of metal to which no rubber is adhering. The common causes of this failure are precure of rubber, low molding temperature, blooming of ingredients and use of incorrect bonding agent.

The fourth is rubber failure, where the strength of rubber is lessthan the bond strength.

All these factors should be considered during the whole process of manufacturing rubber–fabric bonded seals to ensure a quality product.

REFERENCE

1 Gage, F. Rubber to metal bonding. Journal of the IRI 2(1), 1968
2 Ruffel, J.F.E. Elastomer to metal bonding. Presented at West England section Conference at Trowbridge, 5 March, 1968. Journal of the IRI 3(4), 1969
3 Petrie, E.M. Handbook of adhesives and sealants, 2nd edn. McGraw-Hill Professional, 2006

CHAPTER 10

MANUFACTURE OF SEALS AND 'O' RINGS

An outline of the processes used in the manufacture of rubber seals and 'O' rings is given here. Some base formulations of rubber compounds, along with their use and physical property data are also included for reference.

MANUFACTURE OF 'O' RINGS

The specified rubber compound is mixed and calendered and plied up to the desired thickness, then sheeted out on a two-roll rubber mill, or extruded. These unvulcanized rubber blanks are then prepared in for subsequent molding operations one of the following ways:

1. Cutting out a ring from the sheet with a die, or with a knife and template.

2. Cutting a strip from the sheet stock, and butt or bias jointing, to form a ring.

3. Extruding a tube and cutting rings from it.

4. Extruding in cord form, cutting to correct length, and then butt or bias jointing.

5. Wrapping round the calendered sheet on a mandrel to give the required thickness and cutting into rings.

The above methods have the following features and/or drawbacks: Method 3 gives maximum production and minimum rejection; Method 1 is more tedious, and results in more remilling of waste; Method 5 may lead to ply separation in hard stocks; Method 2 is the simplest, but rejections are higher, due to joint separation at the bias cut portion and, with soft stocks, air trapping is expected; Method 4 is prone to lamination faults at the joints.

Vulcanization is done in case hardened mild steel or Nitralloy molds. Large and small 'O' rings can be molded concentrically in the same mold. The flash at the parting line is either stripped off by hand or by trimming in a tumbling barrel using solid carbon dioxide. The rubber is frozen so hard that friction from its motion against the walls of the tumbling barrel removes the flash completely. Large 'O' rings demand hand buffing.

MANUFACTURE OF METAL BONDED OIL SEALS

Metal shells are punched from mild steel plates of the required gauge in a power press. The sides of the disk are raised, and then the center punched out. These three operations can be carried out in a single process using a specially designed die. The rough inner edge of the metal shell is then smoothed out and its the top edge is chamfered in a lathe. The chamfer on the shell not only permits easy assembly of the seal into its housing, but also eases fitting of the shell into the mold.

The shell thus prepared is degreased in hot trichloroethylene, shot blasted and again degreased. The first degreasing prevents the shell becoming contaminated, whilst the second removes any residual grease or dirt adhering to the shell after shot blasting. The bonding agent (usually a mixture of isocynates, chlorinated rubber and phenol formaldehyde resin in fixed proportions) is then applied to the inside of the shell by brushing or spraying. Hand brushing is more successful, though not suitable for higher throughputs. Spraying requires the outside of the shell to be masked. Dipping is an ideal method for high throughputs, however with this method the outside coating of the shell will have to be removed during final finishing of the molded seal, in a centerless grinder.

When the bonding agent has dried the shell is ready for loading with unvulcanized rubber blanks. After filling it is kept in a hydraulic vulcanizing press for curing. After molding, the seals are rough trimmed by cutting off the flash at the metal edge. Removing the flash at the sealing edge is a more delicate procedure.

Thin flashes at the sealing edge are rapidly removed in a lathe chuck, then a fine emery cloth is applied to the edge whilst it is in motion. For seals with a thick flash, a portion is cut away by an attachment which brings up a thin sharp blade at the correct angle. The first method necessitates a perfectly molded sealing edge. The second method is more reliable. The final process is the grinding or turning of the metal case to the required tolerance in a centerless grinder. Finally the garter pin is assembled, then the seal is greased and packed.

Since the seal or 'O' ring are precision made and critical products, extraordinary care has to be exercised in all the manufacturing steps starting with raw material quality control. Setting control limits on tensile strength, hardness, curing characteristics, specific gravity, compression set etc, is essential for maintaining quality control of production batches. Mooney viscosity and its related curves in a rheometer are important measures for process control of raw rubbers and mixed stocks of compound. Mixing the compound and other ingredients is critical since most rejections during calendering, extrusion and molding are due to improper compounding of rubber and mixing.

Neither natural nor synthetic rubbers can be used as they are produced. They must be mixed with other chemicals to get a balance of properties to suit each end-use. In order to determine the exact proportions in phr (parts per hundred rubber) that are needed for a given end use, various aspects of the seals and 'O' rings are taken into consideration, such as working temperature, required oxidation, abrasion and chemical resistance, and functional properties such as oil, fuel, heat and ozone resistance. Operational parameters, such as speed, temperature and pressure in mixing, extrusion, calendering and molding are also important considerations. The quality, durability and dependability of rubber seals have been made possible by the practical science of rubber compounding, as well as design aspects. The rubber technologist is responsible not only for final product quality, but also for the adaptability of a rubber compound to varying manufacturing parameters, such as temperature, pressure and forming in the molding press. It is easy to make a product prototype that has almost any desired properties in the laboratory, but it is not always easy to duplicate this on a large scale, although laboratory data are always valuable for process and product development. Rubber compounding practice has undergone mind-boggling changes in the last few decades, thanks to the multifarious compounding ingredients coming onto the market at such a fast pace, and the sophiscated demands of the engineers who use rubber products. Today, rubber compounding is considered as a science or a technology, where the skilled technologist can predict with a fair degree of accuracy the suitability of an 'O' ring or seal for the ultimate required application.

For proper compounding of a specified compound, a thorough knowledge and understanding of the following points is vital:

1. Selection of the basic type of rubber suitable for the service required.

2. Selection of the process parameters by which the product will be manufactured, i.e. by mixing, extrusion, calendering, molding or forming.

3. It is necessary to know the physical and chemical properties of compounding ingredients while designing a compounds.

4. A comparative study and evaluation of the gum and filled vulcanizates of various types of rubbers would be helpful.

5. An intimate knowledge of the intended end-use.

6. Design aspects, such as wall thicknes, size, and shape of the products and their required mechanical, chemical, and functional properties.

7. Understanding the different uses of rubber compounds for various applications.

SEAL MOLDING SHOP PRODUCTIVITY

Millions of seals and 'O' rings, manufactured from many kinds of rubbers and with mind-boggling varieties of compounding ingredients are produced in order to conform to the varying requirements of the industries that use them. They are designed to optimize production cost, as well as service life. The main constraints faced by the molding shop are the differences in mold output versus press output. The forms depends on the number of cavities and the latter depends on the number of daylights in the hydraulic molding press. The number of operators required to service the presses seriously affects productivity. The most common arrangement in many small and medium factories is to provide as many presses as possible for one person to service. For instance three presses are serviced by one person, a cycle of three presses with respective sets of molds has to be completed within the cure time fixed for one press. The curing cycles have to be fairly short, so that mold cooling and stock scorch are less likely. In this way, a number of presses can be serviced by the same operator.

A modification of, this method is the split cure technique, in which molds are serviced for the second time while the press is closed. For example, a four daylight press is started with the upper two daylights being loaded and curing with serviced molds. Half way through their cure, the lower two daylight are opened and the serviced molds are loaded and the press is closed. A further set of molds comprising two daylight loadings are then charged, and replace those in the top two daylights at the end of their cure. The process is repeated with those molds removed from the upper two daylights being subsequently serviced and used to those in the lower daylights and so on. Similarly, only one daylight out of four need be replaced at each opening of the press, though the press will be opened more frequently during its cycle. In this kind of cyclic operation, the platen utilization is almost 100%, and turnover is huge. However, the curing system has to contain an adequate plateau effect without sacrificing the homogeneity and uniformity of each batch of moldings.

For a press to be operated with efficiency, it should be curing for the maximum possible time, and as much of the area of the platen as possible should be utilized. This has led to the term 'press loading factor' being coined which can be defined as the total area of all molds in the presses divided by the total available platen area. To achieve a high press loading factor, multicavity molds of the same or nearly the same sizes as the platens of the presses must be used. In many cases, there are deviations between specified and the actual ones received by the molds when such high utilization of presses is employed. In a study conducted by production engineers, the acceptable deviation in cure time is 2 to 3 minutes. However, in some cases, the deviation is can be in excess of 20 minutes, in which cases overcuring occurs.

Large differences in cure times are undesirable and should be reduced. Press productivity is associated with molding cycle time and also labor productivity. Ideally, the press cycle consists of the following five stages:

1. Unloading the press

2. Stripping the molds

3. Filling the molds with blanks

4. Loading the press

5. Curing.

If an operator is looking after three presses his cycle of operation will consist of:

1. service press 1 (perform the above operations)

2. service press 2

3. service press 3.

Another way of increasing press utilization is by shortening the molding cycle, by using duplicate molds, so that the stripping and charging operations on one set of molds are being done while the duplicate set is being cured. The use of split cures is, of course, useful since the presses can be curing for as much as 90% of the total cycle time.

BLANK PREPARATION

The manufacturing process for rubber seals and 'O' rings consists essentially of at least five operations:

1. Forming; either calendering, remilling, extrusion or press curing

2. Cutting to produce blanks of the required size from the calendered sheet or extruded profile, or from the remilled sheet for filling the molds

3. Hand building of blanks and fabric inserts

4. Molding and curing

5. Trimming to remove molding flash.

Large factories will have a separate section for blank preparation. A great variety of different cutting methods are employed, chosen according the machines available during production, and the number and type of blanks required. As many as 200–300 different shapes and sizes might be required for production during a week. They all fall into one of the three types:

1. Rectangular blanks cut from milled or calendered sheet.

2. Extruded sections cut to length.

3. Blanks of specific shape cut from milled or calendered sheet.

The type of blank required for a molding may be dictated by its shape or in many cases, by the ease with which each type can be produced with the facilities to hand.

Different blank preparation methods are followed by different factories. They are:

1.   Manual methods: this category includes hand cutting with scissors or knives, hand punching and manually operated cutting presses.

2.   Punching presses, ranging from simple hand-operated fly presses to large power presses.

3.   Circular knives, known as bacon slicers, are used exclusively for cutting extrusions. The knife is water lubricated and the machines fitted with these circular knives are used to cut extruded sections, in general larger than 1 inch (2.5 cm).

4.   Power presses which have different sizes of blades, to which the rubber strip is fed by manual positioning and feeding of the rubber.

5.   Other types of cutters, such as volumetric cutters, for mass production of blanks from the hot rubber that comes directly from the calendar mill or extruder are also in vogue.

TRIMMING OR DEFLASHING

The trimming of rubber moldings, like the preparation of the blanks, is usually carried out in a separate section of the molding department. The speed with which flash can be removed from a molding is dependent not only on its length and thickness, but also upon the shape and position of the parting or flash line, and the method used to remove it. Some trimming methods are suitable only for a limited range of types of flash line. With 'O' rings, the general practice is to position the flash line at a 180° angle where the sealing lip also is positioned. To avoid undue variation in the lip surface because of the flash line, it is sometimes set at a 45° angle.

Various methods are in place for removing molding flash, such as solid carbon dioxide tumbling (where thin flash is frozen and broken during tumbling), hand trimming, buffing, drilling the flashes in bores or inside of rings, hand punching etc. 'O' ring moldings can often be assembled on a mandrel and all buffed at the same time.

Less accessible flash lines are buffed with conical, or other specially shaped, buffing wheels. With low temperature buffing in solid carbon dioxide, the quality of the trim varies considerably from product to product. In such cases, the final finishing is done by hand. Low temperature tumbling is not effective if the flash is more than 0.010 inch (0.03 cm) thick, with about 0.005 inch (0.013) being preferred. The trimming output is always less than the molding output, so a balance has to be reached between these two operations for a continuous and uninterrupted production schedule.

FLUID SEAL RUBBER FORMULATIONS

Many well known seal manufacturers in the USA and UK, and others in Europe, have their own proprietary formulations adapted to the requirements of the aviation, nuclear, mining and automotive industries. Patented formulations are also available for such applications. Such formulations acheive improved resistance of a particular rubber to swelling by fuels and oils by losing low temperature flexibility. If high resistance to oils and flexibility at low temperature are both required, a compromise must be made.

Some workable formulations for use as are appended in the following tables starting points as guidelines for rubber technologists and engineers.

Natural rubber compounds

Rotary seal compound (Table 10.1)

Cure and physical property data

Cure at 297.5°F (147.5°C) (50lb steam pressure) for 30 min

Hardness: 88° to 90° A

Tensile strength: 1820 psi

Breaking elongation: 250%

The heat resistance of this compound can be improved by decreasing the sulfur content and using a higher proportion of a sulfur bearing accelerator, such as tetra methyl thiuram disulfide, but this will result in less resilience and higher compression set.

'O' ring compound (Table 10.2)

Cure and physical property data

Cure at 297.5°F (147.5°C) (50lb steam pressure) for 15 min

Hardness: 63° to 65° A

TABLE 10.1 Rotary seal compound

Ingredients	phr
Smoked sheet	100
Plasticizer	2
Stearic acid	1
Paraffin wax	2
Phenyl-β-betanaphthylamine	1.5
Zinc oxide	5
Medium processing channel black	5
Fine thermal soft black	120
Mercaptobenzothiazole	1.5
Sulfur	2.75
Total	285.75

TABLE 10.2 'O' ring compound

Ingredients	phr
Smoked sheet	100
Plasticizer	2
Stearic acid	1
Paraffin wax	1
Phenyl-β-naphthylamine	1.5
Zinc oxide	5
Medium processing channel black	20
Fine thermal soft black	50
Mercapto benzothiazole	1.25
Sulfur	2.5
Tetramethyl thiuram disulfide	0.15
Total	184.40

Tensile strength: 2600 psi

Breaking elongation: 510%

Modulus at 300% elongation: 1310 psi

Styrene butadiene compounds

Rotary seal compound (Table 10.3)

Cure and physical property data

Cure at 297.5°F (147.5°C) (50lb steam pressure) for 20 min

Hardness: 90° to 95°A

TABLE 10.3 Rotary seal compound

Ingredients	phr
SBR (non-oil extended)	100
Plasticizer	5
Stearic acid	2
Paraffin wax	1
Phenyl-β-naphthylamine	1.5
Zinc oxide	5
Medium processing channel black	75
Fine thermal soft black	120
Dibenzothiazyl disulfide	1.5
Sulfur	2.0
Tetramethyl thiuram disulfide	1.5
Total	314.15

Tensile strength: 1740 psi

Breaking elongation: 330%

Modulus at 200% elongation: 1250 psi

'O' Ring compound (Table 10.4)

Cure and physical property data

Cure at 297.5°F (147.5°C) (50lb steam pressure) for 30 min

Hardness: 57° to 58°A

Tensile strength: 2250 psi

Breaking elongation: 450%

Modulus at 200% elongation: 1750 psi

Butadiene acrylonitrile compounds

In these compounds, different plasticizers may give improved low temperature properties, but many are likely to be leached out in hot oil. PPA (polypropylene adipate) plasticizers are retained, but confer flexibility only at normal temperatures.

Rotary seal compound with high nitrile rubber content (Table 10.5)

Cure and physical property data

Cure at 297.5°F (147.5°C) (50lb steam pressure) for 20 min

Hardness: 77° to 84°A

Tensile strength: 1865 psi

TABLE 10.4 'O' ring compound

Ingredients	phr
SBR (non-oil extended)	100
Process oil	8
Stearic acid	2
Paraffin wax	1
Ethoxytrimethyldihydroquinoline	1.5
Zinc oxide	4
SRF (semi-reinforcing furnace) black	75
Diphenyl guanidine	0.3
Sulfenamide accelerator	1.2
Sulfur	2.0
Total	195.0

TABLE 10.5 Rotary seal compound with high nitrile content rubber

Ingredients	phr
High acrylonitrile nitrile	100
Dibutyl phthalate	10
Stearic acid	1.5
Paraffin wax	2
Phenyl-β-naphthylamine	1
Zinc oxide	5
Medium processing channel black	30
Fine thermal soft black	100
Dibenzothiazyl disulfide	1.5
Sulfur	1.5
Tetramethyl thiuram disulfide	0.15
Total	252.65

Breaking elongation: 455%

Modulus at 200% elongation: 1140 psi

'O' Ring compound with medium acrylonitrile rubber content (Table 10.6)

Cure and physical property data

Cure at 297.5°F (147.5°C) (50lb steam pressure) for 30 min

Hardness: 63° to 64°A

Tensile strength: 1850 psi

Breaking elongation: 440%

One method of counterbalancing the leaching of plasticizer in hot oil is to incorporate natural rubber, or SBR, in a defined proportion. This will swell in oil and counteract the shrinkage caused by the loss of plasticizer.

TABLE 10.6 'O' ring compound with medium acrylonitrile content rubber

Ingredients	phr
Medium acrylonitrile nitrile	100
Di-2-ethyl hexyl phthalate	15
Stearic acid	1.0
Pine tar	5
Phenyl-β-naphthylamine	2
Zinc oxide	5
Fast extrusion furnace black	35
Fine thermal soft black	35
Dibenzothiazyl disulfide	1.0
Sulfur	1.5
Tetramethyl thiuram monosulfide.	0.5
Total	201.0

Rotary seal compound with Nitrile/SBR blend (Table 10.7)

Cure and physical property data

Cure at 297.5°F (147.5°C) (50lb steam pressure) for 20 min

Hardness: 77° to 82°A

Tensile strength: 1760 psi

Breaking elongation: 400%

Modulus at 200% elongation: 1060 psi

If SBR rubber were absent, shrinkage in volume will occur in a hot, low swelling oil due to leaching of dibutyl phthalate. The swelling of SBR compensates for this. In a high swelling hot oil, however, little shrinkage would occur even without SBR, as the rubber will swell more rapidly than the plasticizer is leached. Low swelling oils are the mineral lubricating type, which have a high aniline point (120°C), whereas high swelling oils are fuels, which have a low aniline point (approximately 70°C). The aniline point is defined as the temperature at which equal volumes of aniline and are completely miscible. It is the test oil used in some applications to indicate the aromatic content of oils, and to calculate the approximate heat of combustion. The greater the aniline point, the lower the level of aromatic compounds. A higher aniline point also indicates a higher proportion of paraffin, or saturated hydrocarbons.

There is a relationship between the swelling power of a petroleum-based lubricating or hydraulic oil and some of the parameters normally used to describe it. The most reliable relationship was established by various investigators using either the viscosity-gravity constant or the aniline point of the oil. The range of aniline points investigated covers those normally encountered in automotive lubricating spindle process and hydraulic oils, and it was found that

TABLE 10.7   Rotary seal compound with nitrile/SBR blend

Ingredients	phr
Medium acrylonitrile nitrile	80
Styrene butadiene rubber	20
Dibutyl phthalate	10
Stearic acid	1.0
Phenyl-β-naphthylamine	1
Zinc oxide	5
Paraffin wax	2.0
Medium processing channel black	30
Fine thermal soft black	100
Dibenzothiazyl disulfide	1.5
Sulfur	1.75
Tetramethyl thiuram disulfide	0.15
Total	252.4

the logarithm of the volume increase is inversely proportional to the aniline point of the medium. Commercial grades of neoprene rubbers can therefore be used alongside larger amounts of high aniline point diluents, as they are not expected to swell the rubber to a great extent. Figure 10.1 gives the relationship between the aniline point of an oil and the equilibrium swelling of neoprene vulcanizate containing 20 volumes of semi-reinforcing furnace black.

Chloroprene compounds

Rotary seal compounds (Table 10.8)

Cure and physical property data

Cure at 297.5°F (147.5°C) (50lb steam pressure) for 20 min

Hardness: 84° to 86°A

Tensile strength: 1990 psi

Breaking elongation: 280%

Modulus at 200% elongation: 1530 psi

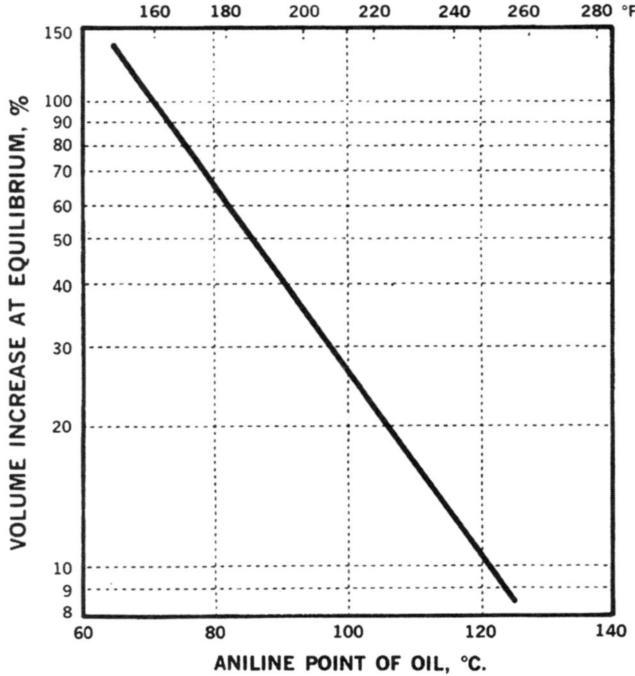

Figure 10.1: Relationship between aniline point and equilibrium swell

TABLE 10.8   Rotary seal compounds

Ingredients	phr
Neoprene WRT grade	100
Light calcined magnesia	4
Phenyl-α-naphthylamine	2
Tri-tolyl phosphate	10
Paraffin wax	2
Zinc oxide	5
Medium processing channel black	35
Fine thermal soft black	100
Stearic acid	1
Ethylene thiourea	0.5
Total	259.5

An alternative formulation using a crystallizing type of chloroprene and a different combination of carbon black is given in Table 10.9.

Cure and physical property data

Cure at 297.5°F (147.5°C) (50lb steam pressure) for 30 min

Hardness: 91° to 93°A

Tensile strength: 1920 psi

Breaking elongation: 150%

Acceleration by ethylene thiourea is more effective with W- and WRT- based compounds than the G types. The neoprene G type compound is cheaper and inferior in abrasion, heat resistance and low temperature properties.

TABLE 10.9   Alternative formulation for rotary seal compounds

Ingredients	phr
Neoprene G grade	100
Process oil	10
Light calcined magnesia	4
Phenyl-α-naphthylamine	2
Paraffin wax	2
Zinc oxide	5
SRF black	60
Medium thermal soft black (MT)	100
Stearic acid	0.5
Ethylene thiourea	0.5
Total	284

'O' ring compounds (Table 10.10)

Cure and physical property data

Cure at 297.5°F (147.5°C) (50lb steam pressure) for 20 min

Hardness: 64° to 65°A

Tensile strength: 2200 psi

Breaking elongation: 320%

Isobutylene-Isoprene (Butyl rubber) compounds

Rotary seal compound (Table 10.11)

Cure and physical property data

Cure at 297.5°F (147.5°C) (50lb steam pressure) for 60 min

Hardness: 89° to 90°A

Tensile strength: 1100 psi

Breaking elongation: 220%

'O' ring compound (Table 10.12)

Cure and physical property data

Cure at 297.5°F (147.5°C) (50lb steam pressure) for 60 min

Hardness: 58° to 60°A

TABLE 10.10 'O' ring compounds

Ingredients	phr
Neoprene WRT	100
Di-2-ethyl hexyl sebacate	6
Di-2-ethyl hexyl phthalate	6
Dark neoprene grade factice	10
Light calcined magnesia	4
Zinc oxide	5
Stearic acid	1.5
Paraffin wax	1
Fine thermal soft black	10
Fast extrusion furnace black (FEF)	35
Phenyl-β-naphthylamine	2
Ethylene thiourea	0.5
Total	181.0

TABLE 10.11   Rotary seal compound

Ingredients	phr
Polysar butyl 301	100
Zinc oxide	5
Stearic acid	2.5
Fast extrusion furnace black	70
Medium thermal black	70
Paraffin wax	2
Tetramethyl thiuram disulfide	1
Dibenzo thiazyl disulfide	0.5
Sulfur	2
Total	253

Tensile strength: 1570 psi

Breaking elongation: 35%

This compound can give a smooth extrusion. Butyl compounds are slow curing.

Brominated butyl compound (Table 10.13)

This compound is faster curing than a normal butyl compound.

Cure and physical property data

Cure at 297.5°F (147.5°C) (50lb steam pressure) for 30 min

Hardness: 73° to 75°A

Tensile strength: 2200 psi

Breaking elongation: 560%

TABLE 10.12   'O'ring compound

Ingredients	phr
Polysar butyl 301	100
Zinc oxide	6
Stearic acid	2.5
Fast extrusion furnace black	70
Process oil	15
Paraffin wax	2
Tetramethyl thiuram disulfide	1
Dibenzo thiazyl disulfide	0.5
Sulfur	3
Total	200

TABLE 10.13   Brominated butyl compound

Ingredients	phr
Hycar 2202 (brominated butyl) rubber	100
Zinc oxide	5
Stearic acid	2.5
Easy processing channel black (EPC)	50
Paraffin wax	2
Tetramethyl thiuram disulfide	1
Dibenzo thiazyl disulfide	0.5
Sulfur	2
Total	163

Silicone rubber compounds

Generally, silicone rubbers are supplied as compounded rubbers ready for molding and curing, but pure gum rubber can be obtained on request from the suppliers. Pure gum requires fillers and vulcanizing agents, usually benzoyl peroxide. According to the pigments used, these rubbers can be either red or white. The pigments are iron oxide and silica, carbon black is generally avoided since it inhibits cure by peroxides. The vulcanization of silicone rubbers differs from normal curing; 3–6 min curing at 130°C is normally needed to set the gum rubber in a mold.

The material is then vulcanized by prolonged exposure to the air. A step cure is used for articles more than 0.125 inch (0.3 cm) thick, e.g. 12 h at 150°C followed by 12 h at 200°C. This prevents porosity in the finished product. The step cure conditions vary according to the grade of silicone rubber and its thickness. Hot air curing must be in a well-ventilated oven, as explosive gases are given off. Bonding of silicone rubbers to metal requires a proprietary bonding agent supplied by the manufacturer. Molding shrinkage of silicone rubbers during cure is high, so molds of precise oversize which take into account the high shrinkage are designed, to achieve correct dimensions after vulcanization.

Poly-acrylic ester compounds

'O' ring compounds (Table 10.14)

Cure and physical property data

   Cure at 297.5°F (147.5°C) (50 lb steam pressure) for 60 min

   Hardness: 60°A

   Tensile strength: 1830 psi

   Breaking elongation: 280%

   Modulus at 200% elongation: 1300 psi

TABLE 10.14   'O' ring compounds

Ingredients	phr
Hycar 4021[*]	100
Fast Extrusion Furance Black	50
Stearic acid	1
Tetramethyl thiuram mono sulfide	1
Trimene base[**]	2
Sulfur	2
Total	156

[*]Hycar 4021: acrylic ester copolymer manufactured by B.F. Goodrich Chemicals, Sp.gr. 1.09, used for 'O' rings and oil seals and gives resistance up to 350°C.
[**]Trimene base: ethyl chloride, formaldehyde and ammonia reaction product manufactured by Uni Royal Chemicals.

No zinc oxide is present in this compound as it decreases the rate of cure. Sulfur is not used as a vulcanizing agent, but a cure modifier, which improves aging. Trimene base is the vulcanizing agent in this rubber. The water resistance of this compound is not good but can be improved by using triethylene tetramine. The trimene/sulfur system is used for maximum heat resistance. Lack of resilience limits these compounds for use in 'O' rings and gaskets.

Polysulfide rubber compounds

Thiokol FA, a polysulfide rubber, does not plasticize on the mixing mill without a chemical plasticizer such as benzothiazyl disulfide. Thikol ST, another grade, has its molecular weight adjusted so that it is millable. Thiokol FA is vulcanized by zinc oxide and Thiokol ST by zinc oxide or a paraquinone dioxime/zinc oxide combination.

'O' ring compounds (Tables 10.15 and 10.16)

Cure and physical property data

Cure at 297.5°F (147.5°C) (50lb steam pressure) for 40 min

Hardness: 60°A

TABLE 10.15   'O' ring compounds with Thiokol FA

Ingredients	phr
Thiokol FA	100
Zinc oxide	10
SRF black	40
Stearic acid	0.5
Diphenyl guanidine	0.1
Dibenzothiazyl disulfide	0.3
Total	150.9

TABLE 10.16 'O' ring compounds with Thiokol ST

Ingredients	phr
Thiokol ST	100
Zinc oxide	0.5
SRF black	60
Stearic acid	3
Paraquinone dioxime	1.5
Total	165

Tensile strength: 1200 psi

Breaking elongation: 570%

Modulus at 300% elongation: 730 psi

Cure and physical property data

Cure at 297.5°F (147.5°C) (50lb steam pressure) for 30 min

Hardness: 70°A

Tensile strength: 1200 psi

Breaking elongation: 280%

Modulus at 200% elongation 875 psi

Under pressure, these Thiokol rubbers distort, although Thiokol ST is superior to FA. However, they are resistant to petroleum solvents, esters and ketones, aromatic fuels, oils, greases and lacquer thinners, ozone, sunlight and ultraviolet light. This makes them very useful for static seals where no other material will serve.

Chlorosulfonated polyethylene compounds (Table 10.17)

In this rubber, the curing system involves magnesia, litharge, or both, or an epoxy resin. Another alternative cure system is tri-basic lead maleate with wood rosin (organic cure).

Relative advantages

Magnesia cure: excellent concentrated acid resistance but low water resistance.

Litharge cure: excellent water resistance but reduced acid resistance.

Organic cure: good water resistance, excellent acid resistance and is non-toxic.

Hypalon 40 is easier to process than Hypalon 20, and is superior also in oil resistance-equal to neoprene in this respect. Hypalon 40 is also superior to Hypalon 20 in hot tear strength.

TABLE 10.17   Chlorosulfonated polyethylene compounds

Ingredients	Magnesia cure phr	Litharge cure phr	Organic cure phr
Hypalon 20 or 40	100	100	100
Magnesia	10	–	–
Litharge	–	25	–
Epoxy resin	–	–	10 to 15
Dibenzothiazyl disulfide	–	0.5	0.5
Diorthotolyl guanidine	–	–	0.25 to 0.5
Dipentamethylene thiuram tetrasulfide (Tetrone A)	2	2	1 to 1.5

Rotary seal compound: organic cure (Table 10.18)

Cure and physical property data

Cure at 297.5°F (147.5°C) (50lb steam pressure) for 40 min

Hardness: 85°A to 86°A

Tensile strength: 2300 psi

Breaking elongation: 150%

'O' ring compound: magnesia cure (Table 10.19)

Cure and physical property data

Cure at 297.5°F (147.5°C) (50lb steam pressure) for 40 min

Hardness: 69°A to 70°A

Tensile strength: 2310 psi

TABLE 10.18   Rotary seal compound: organic cure

Ingredients	phr
Hypalon 20	100
Epoxy resin	15
High abrasion furnace black	55
Dibenzothiazyl disulfide	0.5
Tetrone A	1.5
Polyethylene (low density)	2
Diortho tolyl guanidine	0.25
Total	174.25

TABLE 10.19 'O' ring compound: magnesia cure

Ingredients	phr
Hypalon 40	100
Light calcined magnesia	10
Medium Thermal Black	55
Stearic acid	1
Paraffin wax	2
Tetrone A	2
Process oil	10
Total	180

Breaking elongation: 400%

Modulus at 200% elongation: 950 psi

Butadiene acrylonitrile/polyvinyl chloride blend

Rotary seal compound (Table 10.20)

Cure and physical property data

Cure at 297.5°F (147.5°C) (50lb steam pressure) for 30 min

Hardness: 78°A to 80°A

Tensile strength: 1900 psi

Breaking elongation: 300%

TABLE 10.20 Rotary seal compound

Ingredients	phr
Paracril OZO[*]	100
Di-2-ethyl hexyl phthalate	15
Zinc oxide	5
Stearic acid	1
Pine tar	5
Fine thermal soft black	35
FEF Black	40
Phenyl-β-naphthylamine	2
Sulfur	1.5
Dibenzothiazyl disulfide	1
Tetra methyl thiuram disulfide	0.5
Total	206

[*]Paracril OZO: a 50/50 coprecipitated latex blend of paracril (butadiene acrylonitrile) and PVC resin made as slabs or flakes. Suppliers UniRoyal chemicals.

TABLE 10.21   'O' ring compound

Ingredients	phr
Butakon AC 5502[*]	100
Tri tolyl phosphate	30
Zinc oxide	5
Stearic acid	1
Tetramethyl thiuram disulfide	0.5
GPF (General Purpoise Furnace) black	30
Phenyl-β-naphthylamine	2
Sulfur	1.75
Dibenzothiazyl disulfide	1
Total	171.25

[*]Butakon AC 5502: a blend of acrylonitrile rubber and polyvinyl chloride. ICI product

'O' ring compound (Table 10.21)

Cure and physical property data

Cure at 297.5°F (147.5°C) (50lb steam pressure) for 30 min

Hardness: 63°A to 64°A

Tensile strength: 2400 psi

Breaking elongation: 400%

Modulus at 300% elongation: 1600 psi

The compound will have excellent flame resistance because of the PVC content and the presence of tritolyl phosphate. The rubber can be processed on a cold mill, the PVC content having been gelled during manufacture. Similar compounds can be prepared using a nitrile rubber and incorporating PVC on the mill, but higher mixing temperatures are necessary to get the PVC gelled and fluxed into the nitrile.

Polyurethane compounds

Rotary seal compounds (Table 10.22)

Cure and physical property data

Cure at 287.5°F (147.5°C) (40lb steam pressure) for 60 min

Hardness: 83°A to 84°A

Tensile strength: 4100 psi

Breaking elongation: 320%

TABLE 10.22 Rotary seal compounds

Ingredients	phr
Adiprene C	100
HAF carbon black	50
Coumarone-indene resin	5
Sulfur	0.75
MBTS (dibenzothiazyl disulfide)	4
MBT (mercaptobenzothiazole)	1
Activator RCD 2098	0.35
Total	161.1

'O' ring compounds (Table 10.23)

Cure and physical property data

Cure at 297.5°F (147.5°C) (50lb steam pressure) for 60 min

Hardness: 55°A

Tensile strength: 4200 psi

Breaking elongation: 670%

The rotary seal compound is sticky during mixing, but the 'O' ring compound is not. The activator RCD 2098 is a zinc chloride/MBTS complex. Zinc diethyl dithiocarbamate and other activators containing zinc may also be used, but they are less effective. These polyurethanes have outstanding abrasion resistance, similar to neoprene in ozone and weather resistance, and comparable to nitrile rubbers in oil and fuel resistance. Heat aging is excellent up to 250°F (121°C), and the brittle point is −90°F (−68°C).

Fluorocarbon rubber compounds

The commercial grade Kel F - PCTFE (polychlorotrifluoroethylene) is a fluorocarbon-based polymer copolymerized with vinylidene fluoride. It contains more than 50% fluorine and is

TABLE 10.23 'O' ring compounds

Ingredients	phr
Adiprene C	100
HAF carbon black	20
Di-2-ethyl hexyl phthalate	10
Polyethylene	5
Sulfur	1.5
MBTS	3
MBT	1
Activator RCD 2098	0.35
Total	140.85

commonly abbreviated PCTFE. PCTFE offers the unique combination of non-flammability, chemical resistance, near zero moisture absorption and excellent electrical properties. It also has a useful temperature range of $-400°F$ ($-240°C$) to $+400°F$ ($204°$). PCTFE also has extremely low outgassing, making it well suited for use in aerospace and flight applications. Kel-F is a registered tradename of 3M Company.

Curing systems for Kel-F include organic peroxides, polyamines, polyisocyanates and isocyanate-amine combinations. Benzoyl peroxide is effective and convenient.

'O' ring compound (Table 10.24)

Cure and physical property data

Cure at $297.5°F$ ($147.5°C$) (50lb steam pressure) for 30 min in press, with post cure for 16 h at $300°F$ ($150°C$)

Hardness: $56°A$ to $58°A$

Tensile strength: 3500 psi

Breaking elongation: 500%

Both zinc oxide and lead phosphate act as accelerators. The milling temperature should be $170–190°F$ ($77–88°C$). An after-cure or post-cure is required following press cure. The after curing time depends upon the thickness of the seal cross-section; for thin sections, 1 h at $300°F$ ($150°C$) will be required, for thicknesses up to 0.075 inch (0.19 cm) 5 h and, for thicker sections, 16 h will be required. Initial press cure for benzoyl peroxide stocks can be 15–30 min at $230–300°F$ ($110–150°C$).

Viton A and Viton A-HV grades are from M/s Dupont, USA and they are copolymers of hexafluoropropylene and vinylidene fluoride.

Rotary seal compound (polyamine vulcanization system) (Table 10.25)

Cure and physical property data

Cure at $297.5°F$ ($147.5°C$) (50lb steam pressure) for 30 min in press with post cure for 24 h at $400°F$ ($204°C$). Consecutively, 1 h step cure should also be given at $212°F$ ($100°C$), $250°F$ ($121°C$), $300°F$ ($150°C$) and $350°F$ ($175°C$)

TABLE 10.24   'O' ring compound

Ingredients	phr
Kel-F elastomer 3700	100
Zinc oxide	10
Dibasic lead phosphate	10
Benzoyl peroxide	3
Total	123

TABLE 10.25 Rotary seal compound (polyamine vulcanization system)

Ingredients	phr
Viton A-HV	100
Magnesium oxide	15
Medium thermal carbon black	60
Copper inhibitor 65	2
Diak No1	1.25
Total	178.25

Hardness: 87°A to 88°A

Tensile strength: 2975 psi

Breaking elongation: 120%

Copper inhibitor 65 containing di-salicylal propylene diamine has a retarding effect at processing temperature, but activates cure. Diak No1 (hexamethylene diamine carbamate) is the vulcanizing agent.

Apart from exceptional heat resistance, this compound has high resistance to aliphatic and aromatic hydrocarbons, including carbon tetrachloride and benzene, esters of aromatic acids, and higher ethers (aromatic and aliphatic), and also to phosphoric acid. It is not suitable, however, for low molecular weight aliphatic esters, ketones, and amines including anhydrous ammonia.

STATIC SEALS AGAINST GASES

Permeability of rubbers to gases is a function of solubility and diffusivity. The relative permeability of rubbers to gases is shown in Table 10.26. Permeabilty is a function of the internal structure of the rubber mixture, for instance butyl rubber has low permeability because of the presence of methyl groups, whereas nitrile rubber has low permeability to non-polar gases because of the polar groups in the chain. The higher the nitrile content, the lower the permeability. Below 32°F (0°C), butyl and nitrile rubbers are practically impermeable.

FILLER EFFECTS ON PERMEABILITY OF RUBBERS TO GASES

Depending on the type used, permeability is reduced to about 30% by the addition of 50 phr of carbon black. Addition of fine particle size carbon blacks, however, increases the permeability, although the rate of diffusion is decreased. This can be explained if the gas is absorbed by the carbon black, thereby rendering it immobile. The relationship between permeability and temperature for rubber containing fillers is practically the same as for an unloaded rubber compound. The decrease of permeability to gases due to filler loading is not determined by energy factors, but is physical in nature.

For low permeability, a rubber compound should contain the highest loading of fillers compatible with service requirements, and little plasticizer. The filler is impermeable and the gas must percolate around the particles. Plasticizers separate the rubber chains, thus allowing easier diffusion of gas.

TABLE 10.26   The relative permeability of rubbers to gases

Rubbers	Permeabilty grading
Natural rubber	F
SBR	F
Neoprene	G
Butyl	E
Nitrile copolymers	E
Thiokols	E
Polyurethanes	G-E
Hypalon	G
Silicones	F
Fluorocarbons	G-E
Acrylic ester elastomers	G
Nitrile/PVC blends	E

F: fair; G: good; E: excellent

'O' ring compound for low permeability (Tables 10.27 and 10.28)

TABLE 10.27   'O' ring compound for low permeability

Ingredients	phr
Polysar butyl 200	100
SRF carbon black	30
EPC black	20
Zinc oxide	5
Petroleum jelly	3
MBT	0.5
TMT	1
Sulfur	2
Total	161.5

TABLE 10.28 '**O**' ring compound for low permeability

Ingredients	phr
Polysarkrynac 801(butadiene/acrylonitrile)	100
FEF black	47
Zinc oxide	5
Stearic acid	1
Lightly calcined magnesia	4
MBT	1.5
Sulfur	2
Total	160.5

CHAPTER 11

STORAGE AND SERVICE LIFE OF RUBBER SEALS

As is well known, rubber, has been used by different industries as an engineering material for a wide variety of applications, because of its ability to undergo large deformations without rupture and to recover almost completely when the forces causing such deformations are removed. Because of their chemical structure, rubbers are subject to some chemical degradation and mechanical damage during storage and service. For many applications, such effects are not serious, since suitable maintenance programs are practiced. To reduce the time spent on maintenance, rubber components should be made to function satisfactorily in service, and have adequate storage life. Degradation and mechanical damage during storage and use are related to rubber elasticity and other properties. These are described below.

RUBBER ELASTICITY

Elasticity is exhibited by materials of varying chemical composition. The primary condition for a material to be elastic is that it should predominantly consist of long linear molecules, i.e. it should have a large number of recurring units linked together by strong chemical bonds, and these should be cross-linked. Elasticity is apparently independent of whether these units are identical throughout the whole chain or whether other monomers are present. Properties other than elasticity, however, such heat and fluid resistance, will be markedly affected by the structure and type of the chemical units involved in the chain.

The other requirement essential for elasticity is that there must be a high degree of internal flexibility within the individual chains, together with low intermolecular attraction, or van der Waals' forces, between individual chains. This internal flexibility enables it to take up many extremely irregular configurations, where individual segments move with respect to others. The degree of movement is directly related to temperature.

In a raw or unvulcanized rubber, the individual chains are randomly tangled, but, due to the low van der Waals' forces, the chains can move past each other. This freedom of movement is responsible for the flow which occurs in rubber, whereas the entanglement leads to high internal viscosity, which tends to retain the shape of the rubber mass. Upon the rapid application of a tensile load, there is some degree of alignment of the chains in the direction of application of the load, allowing the rubber to stretch.

Rubber Seals for Fluid and Hydraulic Systems 2010; ISBN: 9780815520757

This, on the basis of the second law of thermodynamics, puts the rubber in a state of decreased entropy, and so the rubber returns to its original state upon release of the load. In other words, rubber has its maximum entropy in the undeformed state, and will maintain this state in the absence of external loads. If, however, the load is applied slowly, or maintained for a long period, slippage of the chains past each other can occur and, on release of the load, the rubber will show some signs of irreversible deformation. In order to restrict the extent of flow under such conditions, intermolecular bonds between adjacent chains, or between segments are introduced and this process is called vulcanization, or cross-linking. This gives an irregular three-dimensional network structure, which tends to restrict the extent of irreversible flow. Because of the distance between cross-links, slippage of the greater part of the chains can still occur and so elasticity is therefore affected very little by normal levels of cross-linking.

VULCANIZATION

As discussed the process of vulcanization involves a chemical modification of the rubber in which cross-links are introduced in a random fashion. Before vulcanization, rubbers are characterized by their viscous state in which elasticity, tensile strength and recovery after deformation are comparatively low. They have a narrow useful temperature range and are soluble in organic solvents. After vulcanization, a highly elastic state is obtained, in which tensile strength and recovery after deformation are high and flow is low. The temperature range in service is extended, and solubility in organic solvents is reduced.

There are various methods of vulcanization, all depending upon the presence, or introduction of reactive sites for the cross-linking reaction. In natural rubber (NR), styrene butadiene (SBR), ethylene–propylene terpolymer (EPDM) butyl and nitrile rubbers, these reactive sites are olefinic double bonds. In other rubbers, the reactive site may be a halogen atom, such as in polychloroprene or fluororubbers, or an active hydrogen, as in silicone rubbers. The process of vulcanization does not entirely eliminate all the viscous element of the rubber properties, although the elastic state is more dominant. The residual viscous element which remains after vulcanization causes some of the undesirable physical properties seen during storage and service of the rubber.

SECOND ORDER TRANSITION, BRITTLE POINT AND CRYSTALLIZATION

Since movements of the individual atoms and segments of the rubber chain are partly due to thermal effects, a lowering of the temperature will tend to slow this down, and the rubber will become progressively stiffer, passing through a leathery stage until it becomes a brittle solid with little extensibility. The temperature at which this change occurs is known as the second order transition temperature, and is regarded as a reference point, below which all rubbers lose their elastic properties, and above which dimensionally stable rubber characteristics prevail. This transition takes place over a temperature range of some 20°C, and is a function of the molecular structure of the rubber. The second order transition is characterized by time effects, important when considering an application in which a rubber seal is required to undergo a fixed deformation in a given length of time. For most applications, the deformation will be large and the time involved short.

At temperatures well above the second order transition temperature, the rate at which the chain segments move is greater than the applied deformation rate and so the rubber deforms satisfactorily. As the temperature is lowered, the molecular deformation rate approaches that of the applied deformation until two rates are equal, at which point the rubber can undergo brittle fracture. This means the rubber has little low-temperature flexibility.

If the deformation and temperature are kept constant but the speed of deformation is increased, again the two rates become equal and brittle fracture occurs. The temperature at which this occurs is known as the brittle temperature. This temperature is not an absolute function of the rubber, but is dependent upon the rate and magnitude of the applied deformation.

In addition to the change in stiffness which occurs with decreasing temperature, there is an additional structural change which occurs in certain rubbers, particularly natural rubber and certain grades of neoprene. On exposure to moderately high temperatures for long periods of time, some segments of the polymer chains, when subjected to stress, can become aligned. In this state the material behaves as a glassy solid, with no elasticity. This phenomenon is known as crystallization, and is dependent on chemical type and structure, and is important during long-term storage of rubber components (see Table 5.4 for details of the stretching crystallization of rubbers).

Like any other engineering materials, rubbers do not remain exactly the same size over large changes in temperature. The coefficient of cubic expansion of rubber is ten times that of steel, so at low temperatures, marked shrinkage can occur, leading to leaks in fluid systems if provision is not made for this in design.

Since all these effects; glass state, strain crystallization and shrinkage, are temperature dependent, the process is completely reversible and the elastic state is restored when the temperature is raised.

EFFECTS OF FLUIDS

Another noteworthy characteristic of rubbers, either natural or synthetic, are their behavior towards fluids. In general, low molecular weight compounds (non-polar) have sharply defined maximum levels that will dissolve in a given fluid. Rubber on the other hand, first swells, absorbing the fluid without true solvation occurring. In the case of non-solvents, this changes the elastic properties very little. In true solvents, however, an increasing amount of fluid is absorbed, leading first to the formation of a gel and finally a true solution. These effects are markedly reduced in the case of vulcanized rubber, so that only the swollen state is attained.

The process of swelling occurs in two stages. Firstly, a rapid uptake of fluids occurs, reaching a fairly well defined equilibrium state, and then secondly fluid is absorbed slowly at an approximately constant rate. This second stage does not take place in the absence of air, and it is therefore assumed that it is related to irreversible breakdown of the rubber. Under most operating conditions, this effect is unlikely, unless there is exposure to air. Prediction of the degree of swelling likely with a given compound in a given fluid is possible using experimental results.

Except in those cases where there is a specific chemical reaction, such as oxidation, chlorination etc. (i.e. action of acids on natural rubber), the action of the fluid is physical, and therefore reversible, so the rubber recovers most of its properties when the fluid is removed.

CREEP AND STRESS RELAXATION

Although the process of vulcanization leads to the elastic state becoming dominant, it does not entirely eliminate the viscous element of rubber properties, so when a constant stress is applied, deformation is not constant, but increases gradually over time. Such an increase is known as creep. Conversely, when a rubber is subjected to a constant strain, a decrease in stress takes place. This behavior is called stress relaxation, or stress decay. Both these behaviors are familiar to engineers in metal springs (see Figures 2.1 and 2.2). At room temperature, these effects are directly related to the viscous element of rubber, and are comparatively small, but, at higher temperatures, chemical processes occur which result in either main chain scission or cross-link scission, so increasing their magnitude.

During use of a rubber seal or 'O' ring, it is impossible to avoid deformation and the resulting effect of stress relaxation. This is a significant subject for study over the service life of the seal. During storage, however, seals should be stored in conditions free from tension, compression or deformation, so avoiding permanent set, which would make the seal unserviceable.

DYNAMIC PROPERTIES – HYSTERESIS

It was mentioned earlier that rubber elasticity is related to the high levels of entropy present in the undeformed state. Deformation, therefore, involves application of energy, some of which is necessary to overcome inter- and intramolecular forces, the remainder is stored, but is released on recovery from the deformation. The recovery stress-strain curve, therefore, never coincides with the loading curve, with the loss of energy being known as hysteresis.

The amount of hysteresis is dependent on the type of rubber and is increased by the presence of filler. At low deformation rates it is very small, but increases rapidly as this increases. Since it is released in the form of heat, high hysteresis is normally not desirable since it leads to a rise in temperature (heat build up), so causing thermal degradation. This phenomenon is very familiar to automobile engineers as, in extreme cases, it can lead to blow-out of a tire.

Since most rubber sealing applications only involve fairly low deformation rates, hysteresis is not likely to have serious consequences. In other applications, where temperatures and rates of deformation are high, the effects could be significant. Hysteresis is, however, important due to its effect on properties such as tearing.

FRICTION, ABRASION AND TEARING

The amount of friction that is generated when a material is pressed against a surface is given by its coefficient of friction, μ. This is the proportion of applied force that is turned into friction, and for normal materials is a constant. This is not the case for rubber, whose coefficient of friction decreases with applied force. This effect is pronounced on smooth surfaces on which the frictional forces eventually become constant, and independent of the load. The value under any given load is dependent on velocity, temperatures, and surface finish of the rubber and contacting surfaces. It can be attributed to two mechanisms, namely adhesion between the rubber and the contacting surfaces, which in turn is dependent on the size of the contacting area, and secondly, mechanical energy losses, which are related to the hysteresis behavior of rubber discussed above. The adhesion behavior depends on the size of the area of contact between the rubber and adjacent surface, which can be increased enormously by surface irregularities. These surface irregularities are due, firstly, to machining the seal housing and, secondly, to the reproduction of the machine marks of the mold on the rubber surface.

It is clear that these local concentrations of stress cause abrasion of rubber which then occurs by detachment of discrete particles. The volume of these is directly related to the size of the contact area. The probability of a particle being torn off on a rough surface is proportional to the elongation at break of the rubber, and therefore on the nature and composition of the rubber compound.

when a dynamic seal starts to operate, two processes occur simultaneously. The first is adhesion between the rubber and the contacting surface, which affects both friction and the modulus of the rubber, and the second is abrasion, which occurs on both rubber and metal. Both adhesion and friction are prevented by lubrication with fluid available from the permissible leak in the system, or while the seal is in contact with fluid during the reciprocal movement of the shaft. In a system with a high shaft speed, most of the abrasion of both rubber and metal, occurs within the first two hours of operation and is virtually negligible thereafter.

A third effect can occur which can be much more serious. This is extrusion particularly common in high-pressure systems, leading to tearing of the rubber at the clearance between housing and shaft. This can be minimized by reducing the clearance as much as possible, and also by the introduction of harder material as a backup seal, or by using a rubber with high hardness.

The tear properties of rubber depend on tear rate and temperature. At low rates, tear has a knotty appearance, whereas tearing at high speeds proceeds steadily, and the torn surfaces are much smoother. In the absence of reinforcement or crystallization, tear behavior is dominated by hysteretic properties in the viscous phase of the rubber. It is related directly to the relaxation of stress that occurs at the tip of the growing tear.

Hysteresis is increased by the inclusion of fillers, but with non-reinforcing fillers, there is a loss of strength due to reduced adhesion between rubber and filler. Stresses on such systems lead to detachment of the filler, so forming vacuoles, which are flaws that may result in premature failure. In the case of reinforcing fillers, a strong adhesion exists between rubber and filler, giving both increased hysteresis and increased strength. This is similar to the effect obtained with rubbers that crystallize on stretching. Improved resistance to tearing is therefore possible by developing compounds that have high hysteresis properties, either by using crystallizing rubbers or by using reinforcing fillers.

THERMAL EFFECTS

No rubber is completely stable with respect to temperature and, at elevated temperatures, all rubbers tend to degrade, mainly by chain scission. Thermal resistance is a function of the polymer structure. As described earlier, stretching a rubber decreases its entropy. As an increase in temperature will increase entropy, strength characteristics are, as a result reduced. With most rubbers, except silicone, there is a reduction of strength of some 60% is near to the upper operational temperature limits for the particular rubber. Conventional heat aging tests on a variety of commercial rubbers show that, if there is a requirement for a long useful life at a temperature in excess of 125°C, only two classes of rubbers; silicones and fluororubbers, need be considered. High temperatures have two types of effect on rubbers; firstly there is an immediate, reversible lowering of mechanical properties, and secondly there is a permanent change, occuring over the longer term. The latter is known as heat aging, and is largely the result of chemical reactions that modify the three-dimensional network structure of the rubber vulcanizate.

Both the immediate and long-term effects of high temperature are important in the performance of rubber seals.

CHEMICAL PROPERTIES

In order to obtain the desired properties in the vulcanized rubber, only a limited amount of cross-linking is needed, usually between 1% and 5%. After vulcanization, therefore, there are a large number of reactive sites that have not been cross-linked and it is these that are prone to chemical attack. The extent to which this may occur is largely dependent on the reactivity of the available sites, and different rubbers vary widely in this behavior. The most common types of chemical attack are those due to oxygen, ozone, light, stress relaxation, heat, and other oxidizing chemicals. If access to the rubber by these agents is prevented, no degradation will occur. This, of course, is impossible in most service conditions of seals, but suitable precautions can be taken during storage to minimize such access, and recommendations for suitable storage of vulcanized rubber products are available for many standard specifications.

Problems caused by rubber degradation occur when a seal is assembled into a component (such as a fuel pump or hydraulic system), especially, when that assembly goes into service.

OXYGEN ATTACK

The more serious cause of deterioration in rubbers is its reaction with atmospheric oxygen. This is possible because rubber is a diene polymer and some, such as natural rubber, EPDM, SBR, nitrile rubber, and butyl rubber, have olefinic double bonds in their structure. Much research work is being done on the oxidative degradation of unvulcanized rubbers, but this is not relevant to the resistance of vulcanized rubbers in storage or in service as their aging behaviors differ widely. Unvulcanized rubber compound has to be vulcanized in order to produce usable products. The nature of the cross-link produced varies considerably, and this can affect the balance of chemical and particularly of physical properties of the vulcanizates.

OZONE ATTACK

Rubber surfaces contains many flaws, where cracks can be initiated via ozone attack. Increasing stress will increase the number of flaws, which leads to a larger number of cracks. The depth of the cracks is inversely related to their number, and so, low stresses that produce long deep cracks are more damaging to rubber than high stresses. Rubbers which do not contain olefinic double bonds, such as silicone and fluorocarbon rubbers, are completely resistant to ozone attack under any conditions of concentration, stress or temperature.

For most sealing applications, attack by ozone is only likely during storage, and this can be reduced by packing the seals properly. With exterior seals, ozone attack is possible, and ozone-resistant rubbers should be selected. If this is not possible, anti-ozonants should be added to the compound.

LIGHT AGING

Light aging is a chemical effect which involves an oxidation process catalyzed by light. This effect mainly occurs during storage, and is prevented by suitable packaging. Light aging can occur in exterior seals, but this can be minimized by the incorporation of carbon black in to the rubber compound. Incidentally, most compounds have carbon blacks by default, and so light colored compounds have to bear with this aging effect!

RESISTANCE TO WATER

Although most rubbers are not much affected by water, under certain conditions polyurethanes and some silicones can be attacked. In the case of polyurethane rubbers, the effect is negligible below about 60°C, but above this temperature, they should be used with suitable precautions, in order to prevent ingress of water, or any particles that are dissolved or suspended in it. With silicone rubbers, attack by water only occurs in the absence of air, hence allowing free access of air will minimize this effect.

COMPRESSION SET

Compression set, which is a measure of the recovery of rubber after release from compressive forces, is a combination of creep and stress relaxation and, since tests are conducted at elevated temperatures, the chemical processes of stress relaxation are likely to predominate. Compression set test conditions have been specified by seal users, and are considered by some engineers to be an excellent indicator of sealing efficiency. Compression tests are used in service specifications as a measure of cross-linking.

STORAGE SPECIFICATION OF SEALS

The defects which occur during storage of seals can be largely prevented by following certain storage and packaging specifications. British standard specification BS ISO 2230:2002 Rubber – Storage – Vulcanized rubber – Vulcanized materials – Packaging – Thermoplastic polymers – Packaging materials – Life (durability) can be referred to in this connection.

FAILURE MODE AND EFFECTS ANALYSIS (FMEA)

Failure mode and effects analysis is a procedure for analyzing of potential failures of seals and 'O' rings within a system of classification by severity, or determination of the effect of failures. It is widely used in manufacturing at various phases of the product life cycle. The causes of failure are any errors or defects in the process, design or item. Effects analysis involves studying the consequences of those failures. Once a leak is traced, Tables 11.1 and 11.2 can be referred to, in order to decide on corresponding counter measures. The FMEA for seals is given in the tables as a general guideline.

PRECAUTIONS WHILE HANDLING 'O' RINGS

1. Do not open the original packing of the 'O' ring until required for installation.

2. Avoid exposure to sunlight and keep in a dry atmosphere. Ultraviolet rays and a damp atmosphere will accelerate aging and lead to dimensional change.

TABLE 11.1 FMEA for static seal 'O' rings

Defects observed	Reason	Solution
Hardening of the seal 'O' ring. If pressed, the seal becomes curved and cracked	Temperature is above the heat resistance capability of the rubber	1 Lower the service temperature 2 Change to a more heat-resistant rubber
Softening of the seal. Becomes soft due to high swelling in fluid, and has changed dimension	The rubber compound is not suitable for the sealing fluid. The seal surface may have some remaining cleaning agent such as light oil, gasoline etc	Reselect the rubber material and eliminate the use of cleaning material.
Loss of elasticity. If the section of the 'O' ring is pressed it deforms leaving a groove	Over compression and high temperature	Reselect the rubber material and redesign the groove dimension
Pressed condition. The outer or inner surface of all or part of the seal is in a pressed condition and damaged	Compression forces exceed the limit and swelling is maximum	Reselect the rubber material for reduced swelling
Crack. If the outer or inner surface of the ring is pressed, the seal 'O' ring cuts off	1 Improper assembly of the seal 2 Compared to the section of the 'O' ring, the groove in the cylinder is shallower than stipulated and was not been assembled properly	Chamfering should be done properly Reselect and redesign the groove dimension
Ozone crack. 'O'ring surface is crimpled	'O' ring is stored in an elongated condition. Affected by ozone	Do not keep the 'O'ring under stretched condition in air Apply a lubricant on the ring surface Avoid exposure to air
Crevice occuring on the inner or outer surface of the 'O' ring because of rubbing	Scratching the outer or inner surface of the 'O' ring during assembly	Use proper tools for assembling
Wear and tear. Abrasion occurred at the contact position	The contact surface with 'O' ring is rough and with change of pressure, abrasion has resulted	Provide a smooth surface in the groove

The above failure mode and effects analysis is given only as an example and cannot be considered as a valid analysis which could be applicable universally, since the operation conditions differ widely from factory to factory.

3. After opening the package, check whether any foreign material is attached to the seal.

4. Do not keep rubber seals around any electrical devices, since electric discharge produces ozone.

5. Do not tie the rings with a thread to hang in a rack.

6. While in storage, dust the rings with dusting powder, such as talc, to protect against attack by oxygen.

When properly stored, the 'O' ring will have a storage life of 10–20 years, depending on the type of rubber.

TABLE 11.2 FMEA for dynamic seal 'O' rings

Defects observed	Reason	Solution
Twist. The 'O' ring is twisted, and deformed, leading to leak	Sliding surface is rough Dimension is not correct Shaft speed is too high Might have been twisted while assembling	Polish the surface Surface finish should be finer Care while assembling
Stuck. Part of the surface of the 'O' ring is stuck and damaged leading to leak	Improper assembly	Use suitable tools for assembling the seals
Abrasion at the periphery	Bad sliding surface Lack of lubrication Ingress of dirt, metal bits and foreign material	Provide smooth surface finish Lubricate properly and fully Remove foreign materials with filters
Loss of elasticity. The 'O' ring is in a pressed condition and deformed in the groove itself	Usage of seal ring with alternating low and high pressures	Provide cooling at the sealing area
Wearing. Part of the sliding surface of the 'O' ring is abraded leading to leakage	Damage on the surface of the sliding part	Provide smooth sliding surface

The above failure mode and effects analysis is given only as an example and cannot be considered as a valid analysis which could be applicable universally, since the operation conditions differ widely from factory to factory.

The lubricant used for assembling the seals should not swell the rubber or lead to leaching. It should not flow like a thin liquid, and its viscosity should not increase at low temperatures. It should not decompose at elevated temperatures, and should stick well to the shaft, so that its film is not removed by the 'O' ring. It should also be compatible with the fluid medium.

BIBLIOGRAPHY

Bailor, F.V. Vamac elastomers – versatilty for auto applications. Rubber World 180, 1979.

Blow, C.M., Exley, K., South Wart, D.W. Penetration of liquids in rubber. J IRI 3(1), 1969.

Brown, R.P., Morrel, S.H., Norman, R.H. Tensile stress relaxation. J Inst Rubber Ind 4, 1973.

Brumback, M.E., Clade, J.A. Airborne Rubber Seals Industrial maintenance. Thomson Delmar Learning, 2003.

Chandrasekaran, V.C. Poly Rev 1(2), 19,1982.

Dibbo, A. The mechanism of vulcanization. Trans Inst Rubber Ind 42, T154, 1966.

Edwards, W.H Unclassified report no. 142. Proceedings of the D.MAT. Conference on Problems in sealing aircraft fluid, May 1967. Directorate of Materials Research and Development, Ministry Of Technology, UK, 1967.

Edwards, W.H. Factors affecting storage and service life of rubber seals. Unclassified report no 142, May 1967. Directorate of Materials Research and Development, Ministry of Technology, UK, 1967.

Fedors, R.F. A method for estimating both the solubility parameters and molar volumes of liquids. Poly Eng Sci 14(2), 147–154, 2004.

Handman, S.E. Piping systems. Kirk–Othmer Encyclopedia of chemical technology, 4th edn, Vol. 9. Wiley-Interscience, 1992.

Hertz, D. Theory of rubber compounding. Educational Symposium, September 24, 25, 1991.

Hertz, D. L. Jr Solubility parameter concepts – a new look. Paper no. 36, presented at a meeting of the Rubber Division, American Chemical Society, Mexico City, May 9–12, 1989.

Holden, G. Elastomers, synthetic – styrene-butadiene rubber. Kirk–Othmer Encyclopedia of chemical technology, 4th edn, Vol. 9. Wiley-Interscience, 1992.

International Programme on Chemical Safety. Concise international chemical assessment document No.23. http://www.inchem.org/documents/cicads/cicads/cicad23.htm# PartNumber:1

King, R.C. Piping hand book. McGraw-Hill Book Co, New York, 1967.

Kroes, M.J., Watkins, W.A., Delp, F. Aircraft maintenance & repair, 6th edn. McGraw-Hill Publishers, Glencoe, 2002.

Lindley, P.B. Engineering design with natural rubber, 4th edn. The Malaysian Rubber Producer's Research Association, London, 1974.

Morton, M. Elastomers, Synthetic. Kirk–Othmer Encyclopedia of chemical technology, 4th edn, Vol. 8, pp 379, 380, 381. Wiley-Interscience, 1992.

National Aeronautical and Space Agency (NASA) web site: http://www.grc.nasa.gov/WWW/ K-12/airplane/thermo.html

National Institute of Industrial Research (NIIR), New Delhi – 110007, India, www. niir.org

Patel, A.O. Elastomers: a literature review with emphasis on oil. RAPRA paper dt. 07.05. 07-08801, 2007.

Pathway Bellows Inc Design Manuals for Rubber Expansion Joints. Pathway Bellows Inc, California, 1978.

Senyek, M. Elastomers, synthetic (polyisoprene). Kirk–Othmer Encyclopedia of chemical technology, 4th edn, Vol. 9. Wiley-Interscience, 1992.

Spoo, B.H. Fluoroelastomers meet new automotive needs. Rubber World 180(1), 44, 1979.

Standards of the EJMA Expansion Joint Manufacturers' Association, New York, 1958.

Stocklin, P. The chemical industry's endeavours to serve the rubber industry. Transact Inst Rubber Ind 42, 118, 1966.

Taylor, J.J. Safety in nuclear power facilities. Kirk–Othmer Encyclopedia of chemical technology, 4th edn, Vol. 9. Wiley-Interscience, 1992.

Wong, C.P. D' Ambra, P. Embedding. Kirk–Othmer Encyclopedia of chemical technology, 4th edn, Vol. 9, pp 379, 380, 381. Wiley-Interscience, 1992.

http://en.wikipedia.org/wiki/Laws_of_thermodynamics

GLOSSARY

Abrasion   progressive wearing away of a surface (like the sealing lip of a shaft seal) by a mechanical action, such as scraping, rubbing or erosion.

Abrasion resistance   resistance of a material to wearing away when in dynamic contact with an abrasive surface.

Additive   material added to a rubber compound to alter its properties, e.g. a reinforcing agent to improve strength, or a plasticizer to aid flexibility and processibility; also known as a filler.

Adhere   to bond together across an area of contact.

Adhesion   ability of rubber, or other material to stick to a contact surface; may result from chemical or physical interlocking.

Adhesive   substance used to hold materials together.

Aniline point   the lowest temperature at which equal volumes of pure aniline and an oil will completely dissolve in one another.

Antirad   a material which inhibits radiation damage.

Atmospheric cracking   cracks produced in the surface of rubber articles that result from exposure to atmospheric conditions.

Backup ring   anti-extrusion device. A ring of relatively hard and tough material placed in the gland between the 'O' ring and groove side walls, to prevent extrusion.

Beta particles   negatively charged particles (or electrons), characterized by limited penetration.

Blisters   raised spots in the surface, or a separation between layers that forms a void or air-filled space, in a vulcanized article.

Bond   adhesion between a shaft seal's rubber sealing lip and the metal case. The term is commonly used to denote the attachment of a given rubber to some other member. Bonds may be classified by type as follows: (a) mechanical bond: purely physical attachment accomplished by such means as 'through' holes interlocking fingers, envelope design, riveting etc; (b) 'cold' bond: adhesion of previously vulcanized rubber to another member through use of suitable contact cements; (c) 'vulcanized' bond: adhesion of a rubber to a previously primed surface using heat and pressure, thus vulcanizing the rubber at the same time.

Bonded seal   a type of shaft seal with a rubber sealing element that has been bonded to a metal disk during molding.

Cavity   hollow space within a mold in which uncured rubber is loaded, shaped and vulcanized; also known as a mold cavity.

Chamfer   beveled edge of a component that facilitates assembly of a seal onto a rod or shaft, or into a cylinder or housing.

Clearance gap   gap between two mating surfaces, such as the necessary gap between a moving shaft and the housing in which it moves. A shaft seal can block this gap and prevent lubricant leakage.

Coefficient of thermal expansion   may be linear or volumetric: (a) the coefficient of linear thermal expansion is the change in length measured for each unit of length for a one degree rise in temperature; and (b) the coefficient of volumetric thermal expansion is the change in volume divided by the product of the original volume and the change in temperature. The coefficient of volumetric thermal expansion is three times the coefficient of linear thermal expansion for a solid material.

Compression set   (a) the amount, expressed as a percentage of deflection, by which a rubber specimen differs from its original thickness after release of a compressive load; and (b) the end result of a progressive stress relaxation. In terms of the life of a shaft seal, stress relaxation is the process of elongation, whereas compression set is the result.

Copolymer   a polymer consisting of two different monomers that are chemically combined.

Crack   sharp break or fissure in a rubber surface caused by excessive strain, and/or exposure to detrimental environmental conditions, such as ozone, weather or ultraviolet (UV) light.

Cure   heat-induced process whereby the long chains of the rubber molecules become cross-linked by a vulcanizing agent, and form three-dimensional elastic structures. This reaction transforms soft, weak, non-cross-linked materials into strong elastic products. It is also known as vulcanization.

Cut   slice-like opening in a rubber surface caused by unwanted contact between the surface and a sharp object.

Cylinder   chamber in which a piston, plunger, ram, rod or shaft is driven by or against the system fluid.

Deformation   change in the shape of a seal, or seal component, as a result of compression; also known as deflection.

Differential thermal expansion   variance in the heat-induced rates of expansion for two different materials (such as the metal of a housing bore and the metal case of a shaft seal); this variance may lead to the formation of a gap between the two and this may allow leakage.

Diffusion   the mixing of two or more substances (solids, liquids, gases or combinations thereof) due to the intermingling motion of their individual molecules. Gases diffuse more readily than liquids, similarly, liquids diffuse more readily than solids.

Dynamic seal   seal functioning in an environment in which there is relative motion (e.g. rotary, reciprocating or oscillating) between the mating surfaces being sealed.

Eccentricity   the variation of the shaft surface with reference to the centerline of the shaft.

Elastomer   any natural or synthetic material meeting the following requirements: (a) it must not break when stretched to 100%; and (b) after being held at 100% stretch for 5 minutes then released, it must return to within 10% of its original length within 5 minutes.

Elasticity   a rubber's inherent ability to readily regain its original size and shape after being released from a deforming load.

Elongation   percentage increase in original length (strain) of a specimen produced by a tensile force (stress) applied to it. 'Ultimate elongation' is the elongation at the moment that the specimen breaks.

Extrusion   distortion or flow, under pressure, of a portion of a seal into clearance between mating metal parts.

Evaporation   the conversion from liquid to vapor state of a fluid.

Fatigue resistance   the ability to withstand fatigue caused by repeated bending, extension or compression; also known as flex resistance.

Film thickness   in a shaft seal, the small distance between the sealing lip and the shaft that is typically occupied by a thin film of lubricant.

Flash   excess rubber remaining on the parting line of a molded product.

Flex resistance   the ability to withstand fatigue caused by repeated bending, extension or compression; also known as fatigue resistance.

Garter spring   a helically coiled spring, typically made of carbon steel or stainless steel wire, formed into a ring, and used in a shaft seal to help maintain contact between the sealing lip and the shaft.

Hardness   measure of a rubber's relative resistance to an indenter point on a testing device. Shore A durometers gauge soft to hard rubber. Shore D durometers are more accurate for samples of ebonite.

Heat resistance   a rubber compound's ability to undergo exposure to some specified level of elevated temperature, and still retain a high level of its original properties; also known as heat aging or air aging.

Housing   a cylindrical surface machined on the inside to mate with the outside diameter of a shaft seal; also known simply as the bore.

Interference   difference between the diameter of a shaft seal's sealing lip and that of the shaft to be sealed; interference is designed so that the lip diameter is smaller than the shaft diameter, thus ensuring the formation (and maintenance) of a contact point between the lip and the shaft.

Lip seal   device utilizing the planned interference between a rubber lip and a mating surface (such as a shaft) to prevent leakage.

Load   actual pressure at a sealing face; in the case of a shaft seal, the sum of the elastomeric lip's inherent beam force, the hoop force (as a result of lip stretch upon installation) and the garter spring tension, all of which contribute to shaft loading at the contact point.

Low temperature flexibility   the ability of an elastomeric product (such as the sealing lip of a shaft seal) to resist cracking or breaking, when flexed or bent at low temperatures.

Mating surfaces   points where different parts of an assembly meet.

Memory   a rubber's ability to regain its original size and shape following deformation during its raw and vulcanized state.

Modulus   the force (stress) in psi (pounds per square inch) required to produce a certain elongation (strain), usually 100%, in a material sample; a good indication of toughness; also known as tensile modulus or tensile stress, but not the same as the shear modulus.

Molded lip seal   a shaft seal with a sealing lip formed by molding, rather than by trimming with a knife.

Mooney scorch   measurement of the rate at which a rubber compound will cure, or set up by means of the Mooney viscometer test instrument.

Mooney viscosity   measurement of the plasticity or viscosity of an uncompounded or compounded, unvulcanized rubber seal material, by means of the Mooney shearing disk viscometer.

'O' ring   a circular item with round cross-section which effects a seal through squeeze and pressure.

Oil seal   a specific type of shaft seal designed to retain oil.

Optimum cure state of vulcanization at which the most desirable combination of properties is attained.

Over-cure a degree of cure greater than the optimum causing some desirable properties to be degraded.

Permanent set the deformation remaining after a specimen has been stressed in tension for a definite period, and then released for a definite period.

Permeability the rate at which a liquid or gas under pressure passes through a solid material by diffusion and solution. In rubber terminology, it is the rate of gas flow expressed in atmospheric cubic centimeters per second through a rubber material one centimeter square and one centimeter thick (atm cc/cm^2/cm sec).

Post cure the second step in the vulcanization process for more sophisticated rubbers, such as Viton. Provides stabilization of parts, and drives off decomposition products resulting from the vulcanization process.

Radiaton electromagnetic, wave-like emissions such as gamma, beta and ultraviolet rays etc.

Radiation damage a measure of the loss of certain physical properties of organic substances, such as elastomers, due principally to ionization of the long chain molecule. It is believed that this ionization process (i.e. electron loss) results in redundant cross-linking and possible scission of the molecule. This effect is cumulative.

Radiation dosage the total amount of radiation energy absorbed by a substance. This value is usually expressed in ergs per gram and is denoted by the following units: (a) Roentgen: a quantity of gamma or x-ray radiation equal to approximately 83 ergs of absorbed energy per gram of air. (b) REP (Roentgen equivalent-physical): a quantity of ionizing radiation that causes an energy absorption of approximately 83 to 93 ergs per gram of tissue. (c) REM (Roentgen equivalent-man): similar to REP except used to denote biological effects. (d) RAD: the unit of dosage related to elastomers. It is independent of type of radiation or specimen and denotes an energy absorption level of 100 ergs per gram (of elastomer). The RAD is approximately equal to 1.2 Roentgens.

Reciprocating seal a dynamic seal used to seal pistons or rods that are in linear motion.

Register, off or uneven material dispersed in an elastomer to improve compression, shear or other stress properties.

Resilience a compound's ability to rapidly regain its original size and shape following deformation; also known as rebound.

Rheometer a cure meter which determines and plots a cure curve, illustrating the state of cure for a given time and temperature; typically either an oscillating disk rheometer (ODR) or a moving die rheometer (MDR).

Roughness   closely-spaced irregularities on a shaft surface that are the result of manufacturing and/or cutting, as by tools or abrasive materials.

Scorching   the premature curing of rubber during storage or processing, usually caused by excessive heat.

Seal cocking   misalignment of a shaft seal such that it is not perpendicular to the bore in which it is supposed to fit, and the shaft it is supposed to seal; may be caused by incorrect installation or improper design; also known simply as cocking.

Shaft   rotating, reciprocating or oscillating component that operates within a cylinder or housing.

Shaft diameter   diameter of the shaft expressed in inches or millimeters.

Shaft finish   usually meant to be the surface roughness measured in microinches; a low finish number is indicative of a smoother surface than a high finish number; rotating shafts need to be finished in accordance with the RMA (Rubber Manufacturers Association) specifications; also known as surface finish.

Shaft seal   a dynamic seal designed to retain or contain fluids, and/or exclude foreign materials through the exertion of radial pressure (due to interference) on a moving shaft; also known as oil seal or radial lip seal.

Shelf aging   the change in a material's properties that occurs over time in storage.

Shrinkage   (a) decreased volume of seal, usually caused by extraction of soluble constituents by fluids followed by air drying; (b) difference between finished part dimensions and mold cavity used to make the part.

Spring-loaded   used to describe a shaft seal that has a garter spring as part of its sealing lip.

Static seal   seal functioning in an environment in which there is no relative motion between the mating surfaces being sealed.

Stress relieving   process of relieving stresses in an unassembled coiled spring through exposure to heat; intended to help ensure that the spring force will not be adversely affected by heat during actual service.

Squeeze   cross-section diametral compression of an 'O' ring between the surface of the groove bottom and the surface of an other mating metal part in the gland assembly.

Strain   deflection due to a force.

Stress   force per unit of original cross-section area.

Swell   increased volume of a specimen caused by immersion in a fluid (usually a liquid).

Tear   a separation or pulling away of part of a sealing structure.

Tear resistance   resistance to the growth of a nick or cut in a rubber specimen when tension is applied.

Tensile strength   force in pounds per square inch (psi) required to break a rubber specimen.

Threshold   the maximum tolerance of a rubber to radiation dosage expressed as a total number of ergs per gram (or rads) beyond which the physical properties are significantly degraded. This is generally an arbitrary value, that depends on function and environment.

Under-cure   degree of cure that is less than optimum. May be evidenced by tackiness, loginess or inferior physical properties.

Viscometer   shearing disk device used to gauge the viscosity of a rubber sample under heat and pressure. Often referred to as the Mooney viscometer, this device was once the most common tool for determining processing characteristics, but has now largely been replaced by the rheometer.

Viscosity   resistance to flow; the thicker the substance (such as a liquid), the more viscous it is, i.e. the less it flows.

Vulcanization   the heat-induced process whereby the long chains of the rubber molecules become cross-linked by a vulcanizing agent to form three-dimensional elastic structures. This reaction transforms soft, weak, non-cross-linked materials into strong elastic products; also known as cure.

stimulation fluid, 46–7
stretching crystallization, 50–1
well fluid, 45
Oil-resistant elastomers, 81
Oil-resistant synthetic rubbers
and polymerization type, 92
Oxygen, 11, 85
attack, 135
Oil seals, 18–20
Ozone attack, 136
'O'rings, 7, 12–13, 37–8
compounds design for, 57
cross-section, 14–15
for rotary sealing application,
16–18
precautions, handling, 138–9
reciprocating applications, 13–14
static application, 13
'O' rings, manufacture of, 103
blank preparation, 107–8
filler effects, on permeability of
rubbers, 128
fluid seal rubber formulations,
109
brominated butyl compound,
117–18
butadiene acrylonitrile
compounds, 112–14
butadiene acrylonitrile/
polyvinyl chloride blend,
123–4
chloroprene compounds,
114–17
chlorosulfonated polyethylene
compounds, 121–2
fluorocarbon rubber
compounds, 125–7
isobutylene–isoprene (butyl
rubber) compounds, 117
natural rubber compounds,
109–10
poly-acrylic ester compounds,
119
polysulfide rubber compounds,
120–1
polyurethane compounds,
124–5
silicone rubber compounds,
118–19
styrene butadiene compounds,
110–11

seal molding shop productivity,
106–7
static seals against gases, 127–8
trimming/deflashing, 108–9

P

Packings, 37
Permanent set, 8
Peroxide vulcanization, 62
Phenol formaldehyde (PF) resins,
60, 96
Phenolic resins, 60
Poly-acrylic ester compounds, 119
Polybutadiene rubber (BR), 60
advantages and limitations, 90
Polyisocyanates, 96
Polyisoprene rubber
advantages and limitations, 90
Polysulfide rubber compounds,
41, 120
'O' ring compound, 120–1
rotary seal compound, 120–1
Polytetrafluoroethylene (PTFE), 97
Polyurethane compounds,
124–5
advantages and limitations, 88
'O' ring compound, 125
rotary seal compound, 124–5
Polyvinyl chloride (PVC), 60–1
PPA (polypropylene adipate), 112
Pre-cross-linked butyl rubber, 63
Properties of rubber, for seal
functional requirements,
7–10
fluid resistance, 11–12
high temperature behavior, 11
incompressibility, 12
low temperature behavior, 10–11
mechanical seals, 21–2
oil seals, 18–20
'O'rings, 12–18
sealing lip design, 20
stretching, 11
Pump assemblies in nuclear
plants, seals in, 34–5

R

Rad, 24
Radiation, cross-linking by
chemical mechanism of, 27–9

Radiation units, 23–5
Radiation-resistant rubber seals,
26–7
Rated movements, 78
Reclaimed rubber, 70
Reinforcing fillers, 59, 60
Rem (radiation equivalent man), 24
Roentgen, 23
Rotary seal compound
brominated butyl compound,
117–18
butadiene acrylonitrile
compounds 112–14
butadiene acrylonitrile/polyvinyl
chloride blend, 123–4
chloroprene compounds,
114–16
chlorosulfonated polyethylene
compounds, 121–2
fluorocarbon rubber compounds,
125–7
isobutylene–isoprene (butyl
rubber) compounds, 117
natural rubber compounds,
109–10
poly-acrylic ester compounds,
119
polysulfide rubber compounds,
120–1
polyurethane compounds, 124–5
silicone rubber compounds,
118–19
styrene butadiene compounds,
110–11
Rotary shaft seals, 29
Rubber expansion joints, 71, 74
advantages, 75–6
in chemical process industry, 73
constructional features, 76–8
expansion and compression
strains, 76
in food and beverages
industry, 73
in heating and air conditioning
systems, 74
in hydrocarbon process
industry, 74
manufacture of, 78–9
multiple bellows in industrial
plants, 74
Rubber failure, 101